D0078840

Titles in Greenwood Guides to Great Ideas in Science

Brian Baigrie, Series Editor

Electricity and Magnetism: A Historical Perspective
Brian Baigrie

Evolution: A Historical Perspective
Bryson Brown

The Chemical Element: A Historical Perspective
Andrew Ede

The Gene: A Historical Perspective
Ted Everson

The Cosmos: A Historical Perspective
Craig G. Fraser

Planetary Motions: A Historical Perspective
Norriss S. Hetherington

Heat and Thermodynamics: A Historical Perspective
Christopher J. T. Lewis

Quantum Mechanics: A Historical Perspective
Kent A. Peacock

Forces in Physics: A Historical Perspective
Steven Shor

EVOLUTION

EVOLUTION

A Historical Perspective

Bryson Brown

Greenwood Guides to Great Ideas in Science
Brian Baigrie, Series Editor

GREENWOOD PRESS
Westport, Connecticut • London

Library of Congress Cataloging-in-Publication Data

Brown, Bryson.
    Evolution : a historical perspective / by Bryson Brown.
        p. cm. — (Greenwood guides to great ideas in science, ISSN 1559–5374)
    Includes bibliographical references and index.
    ISBN-13: 978–0–313–33461–0 (alk. paper)
    1. Evolution—History.   I. Title.
    QH361.B76   2007
    576.8—dc22   2007024669

British Library Cataloguing in Publication Data is available.

Library of Congress Catalog Card Number: 2007024669
ISBN-13: 978–0–313–33461–0
ISSN: 1559–5374

First published in 2007

Greenwood Press, 88 Post Road West, Westport, CT 06881
An imprint of Greenwood Publishing Group, Inc.
www.greenwood.com

Printed in the United States of America

The paper used in this book complies with the
Permanent Paper Standard issued by the National
Information Standards Organization (Z39.48–1984).

10   9   8   7   6   5   4   3   2   1

For Kalen,
Goodnight, sweet Prince...

CONTENTS

LIST OF ILLUSTRATIONS

SERIES FOREWORD

The volumes in this series are devoted to concepts that are fundamental to different branches of the natural sciences—the gene, the quantum, geological cycles, planetary motion, evolution, the cosmos, and forces in nature, to name just a few. Although these volumes focus on the historical development of scientific ideas, the underlying hope of this series is that the reader will gain a deeper understanding of the process and spirit of scientific practice. In particular, in an age in which students and the public have been caught up in debates about controversial scientific ideas, it is hoped that readers of these volumes will better appreciate the provisional character of scientific truths by discovering the manner in which these truths were established.

The history of science as a distinctive field of inquiry can be traced to the early seventeenth century when scientists began to compose histories of their own fields. As early as 1601, the astronomer and mathematician Johannes Kepler composed a rich account of the use of hypotheses in astronomy. During the ensuing three centuries, these histories were increasingly integrated into elementary textbooks, the chief purpose of which was to pinpoint the dates of discoveries as a way of stamping out all too frequent propriety disputes and to highlight the errors of predecessors and contemporaries. Indeed, historical introductions in scientific textbooks continued to be common well into the twentieth century. Scientists also increasingly wrote histories of their disciplines—separate from those that appeared in textbooks—to explain to a broad popular audience the basic concepts of their science.

The history of science remained under the auspices of scientists until the establishment of the field as a distinct professional activity in the middle of the twentieth century. As academic historians assumed control of history of science writing, they expended enormous energies in the attempt to forge a distinct and autonomous discipline. The result of this struggle to position the history of science as an intellectual endeavor that was valuable in its own right, and not merely in consequence of its ties to science, was that historical studies of the natural sciences were no longer composed with an eye toward educating a wide audience that included nonscientists but instead were composed with the aim of being consumed by other professional historians of science. And as historical breadth was sacrificed for technical detail, the literature became increasingly daunting in its technical detail. While this scholarly work increased our understanding of the nature of science, the technical demands imposed on the reader had the unfortunate consequence of leaving behind the general reader.

As Series Editor, my ambition for these volumes is that they will combine the best of these two types of writing about the history of science. In step with the general introductions that we associate with historical writing by scientists, the purpose of these volumes is educational—they have been authored with the aim of making these concepts accessible to students—high school, college, and university—and to the general public. However, the scholars who have written these volumes are not only able to impart genuine enthusiasm for the science discussed in the volumes of this series, they can use the research and analytic skills that are the staples of any professional historian and philosopher of science to trace the development of these fundamental concepts. My hope is that a reader of these volumes will share some of the excitement of these scholars—for both science and its history.

Brian Baigrie
University of Toronto
Series Editor

ACKNOWLEDGMENTS

This project grew out of a course on the world view of the earth and life sciences that Ron Yoshida and I developed in the late 1980s. Ron and I taught the course together until his retirement in 1996. During that time I focused on the earth sciences, while Ron's part of the course was an introduction to the philosophy of biology, and especially evolution. His clear and careful approach to the issues was the model for my solo version of the course, which continues to be an important part of my teaching here in Lethbridge. In the book I have tried to keep the focus on the science—but at several points philosophical concerns come to the fore. My aim, when addressing those issues, has been to say just enough to clear the air, and, in particular, to reduce the temptations of selective skepticism, to counter naïve views about the nature of science and evidence, and to calm fears that a moral vacuum lurks at the heart of a fully natural view of the natural world.

I thank Greenwood Press for the opportunity to present, in this volume, a short historical account of the central idea of biology, which I have been thinking, reading, and teaching about all these years. I also thank the series editor, Brian Baigrie, for inviting me to undertake the adventure of writing a serious book for a non-professional audience.

Three student assistants have contributed to the project—early on, my daughter, Tessa, gathered and organized material for the book. David Boutland did the same, and began the search for images of sufficient quality and interest to satisfy Kevin Downing at Greenwood Press. Finally, Mark Huff completed the image search with great success (due entirely to his own organizational skills, persistence, and plain hard work). I thank them all. Of course they are entirely free of responsibility for any errors and infelicities that remain. For these, I am solely responsible.

INTRODUCTION: LIFE AND THE PAST

NATURAL SELECTION

Nothing in biology makes sense except in the light of evolution.
                                        —Theodozius Dobzhansky

Darwin's theory of evolution by natural selection is one of the great scientific ideas of the nineteenth century; as Dobzhansky's famous essay argues, evolution is the central theme in all that we now know about life on earth. But Darwin's ideas, the modern synthesis of Darwin's natural selection with genetics, and even the outlines of the biochemical revolution that began in the mid-twentieth century, are completely accessible. Unlike modern physics, a very good understanding of evolution can be had without a prior course in higher mathematics. Understanding evolution does not demand a mastery of strange and intricate ideas about space and time, or quantum mechanical measurements and non-local interactions. It requires nothing more than common sense and a little thought. Evolution can be explained in plain language, and is supported by straightforward observations about living things.

There are three key premises in Darwin's argument for natural selection. The first premise is competition: not every organism can survive and reproduce to its full potential. This is pretty obvious when you think about it, given how many offspring trees, insects, or human beings are capable of producing, and the limited resources these organisms depend on to grow and reproduce, not to mention the predators that aim to turn them into food. The upshot is that, sooner or later, there is competition for space, or food, or mates, or to escape the lion's jaws. In short, there is competition for all the resources that begin to run short as the population grows. As Darwin put it,

> We behold the face of nature bright with gladness, we often see superabundance of food; we do not see, or we forget, that the birds which are idly singing round us mostly live on insects or seeds, and are thus constantly destroying life; or we

forget how largely these songsters, or their eggs, or their nestlings are destroyed by birds and beasts of prey; we do not always bear in mind, that though food may be now superabundant, it is not so at all seasons of each recurring year.

Darwin, (1859, 62)

The second premise is heritable variation. Individual organisms vary widely, even when they are members of the same species. One maple tree may grow straight and tall, while another is bushier. One osprey may be just a little faster in the air, while another has stronger talons. Our success in breeding crops to be more productive, chickens and cattle to put on weight faster, and dogs in all their variety, bigger or smaller, longer or shorter haired, floppy or pointy eared, shows that some of these variations are *heritable:* the descendents of ancestors with a certain trait are more likely to have that trait themselves.

The third premise connects the first two: some of the heritable variations make a difference in the competition. Perhaps the taller maple will overshadow its bushier competitor, winning better access to the sunlight it needs to nourish itself. Perhaps the osprey with stronger talons will be better at plucking fish from the water, feeding its offspring successfully when the faster osprey cannot. It doesn't matter what the advantage is or how it works. So long as there are heritable differences between individual organisms that make some of them more likely to survive and reproduce than others, evolution will occur.

When individuals with helpful heritable variations out-perform other organisms by surviving and reproducing more successfully, those helpful variations become more common in the next generation. The population of organisms has changed—it has *evolved.* Of course this won't go very far without a source of new variation. But any process that produces changes, good, bad, or indifferent, in what an organism's parent(s) pass on to it can provide the new variations. So long as reproduction doesn't always make perfect copies of the parents, there will be changes, and selection will operate on them, eliminating the bad, ignoring the indifferent, and spreading the good through the population over time.

This is no more than common sense. But we do need to be careful about the language we use: we must not confuse this *technical* use of the words *good, bad,* and *indifferent,* with the more familiar sense of these words (roughly, being *worthy* of approval, disapproval, or indifference). Variations are good, bad, and indifferent by these standards *if* and *only if* they improve, reduce, or leave unchanged the likelihood that an organism with that variation will survive and reproduce. It seems strange to say that a variation that makes some *Escherichia coli* bacteria survive and reproduce more vigorously than other *E coli,* and thereby causes many people to get sick or even die, is *good.* But in our technical sense, it *is* good (for those bacteria, in those circumstances): natural selection will favor such a variation if it comes along.

The fact that natural selection is perfect common sense brings us to a broader point about science. Science in general is common sense *extended.* It builds on our common sense understanding of the world, and it extends and improves our common sense ideas about how to find and figure things out, our methods for investigating the world. Scientists systematically develop new ideas about the world, along with new methods, technology, and procedures for testing and improving them. The result, once you come to understand it and see how it works, is that the scientific world view is extremely convincing. It is the product of a long effort to find unified explanations for a rich body of well-established evidence. Scientists continually criticize and improve on the views proposed so far, testing them with new observations, extending their theories and identifying new implications. The main aim of this brief history of evolution is to show how this process of observing, testing, refining, and revising has convinced biologists that life on earth descended from a single common ancestor, and that natural selection is one of the key processes responsible for producing, from that small beginning, the amazing variety of life we find on earth today.

But before we begin that journey, I have an important philosophical job to do. It will be hard to persuade you that the evidence for evolution is good, if you think that we really can't know anything about the past at all, or (as many creationists have claimed) that any conclusions scientists draw about *unwitnessed* past events are merely theoretical. It's also important to understand that, at least in science, theory does not mean mere speculation or something that we have no evidence for. Scientific theories must really do some work for us. So we'll end this introduction with an examination of *skepticism about the past*. A skeptic about the past is someone who claims that we simply can't know anything about the past. After all, the skeptics urge, the past cannot be observed, because anything we observe must be in the here and now.

The point I want to make is simple and straightforward: we have lots of evidence for claims about the past. But if we begin by setting all that evidence aside as suspect, then the skeptic wins. That is, if we surrender all the links between things we observe now and our conclusions about the past, then what's left can't establish anything about the past. On the other hand, if we begin by considering how evidence we gather in the present tells us things about the past, and why that evidence is convincing, then skepticism about the past is no more tempting than any other form of radical skepticism. Knowledge of the past depends on understanding the processes that produced the evidence we now observe. But our ideas about these processes are not arbitrary or untested. We test them by examining what they tell us about the past and what that implies about the evidence we can gather now. The evidence and processes used in science in general, and biology in particular, have passed these tests with flying colors.

SKEPTICISM ABOUT THE PAST

There are all kinds of things we claim to know about the past. If these claims are right, we must have some evidence that *justifies* the things we claim to know. Consider some things I know about the past:

- I had a bagel for breakfast this morning.
- John F. Kennedy was assassinated in 1963.

- Aristotle taught Alexander the Great.
- The Sumerians developed the cuneiform system of writing.
- Glaciers spread across northern Europe and North America within the last 20,000 years.
- Trilobites were arthropods that thrived during the Paleozoic era.

How do I justify these claims? The answers are obvious at first—I remember eating the bagel, and I remember both JFK's assassination and that it took place during the year I was in second grade, that is, in 1963. I've read about Aristotle and his place in the Macedonian court—this is well documented, a settled fact about ancient history. I've also read about cuneiform writing and its roots in simple accounting techniques, in a scientific journal. I've directly seen evidence of glaciation (striations, polished pavements, kettle lakes, moraines, drumlins, etc.) and I've read a lot about the evidence, its history, and the conclusions geologists have drawn from it. Finally, I've seen fossils of trilobites and I've read about them as well. Their place in the geological column is securely settled: their fossils only occur in Paleozoic strata. Finally, we know that they were arthropods because they have certain features that are shared with, and only with, other arthropods.

Someone defending general skepticism about the past will look for something in common amongst these justifications, some *vulnerability* she can identify that they all share. And they do all have something in common: they all depend on assumptions about *processes* linking facts about the present situation to some fact(s) about the past.

When I talk about what I remember, I assume that some process connects my present memory to the facts I am remembering. Something happened (I ate a bagel; JFK was assassinated) and I was or became aware of it (my senses detected it, in the first case; I heard it from a friend, who heard it over the radio in the second). That awareness left a persistent trace (a memory) that remains in me today, a trace that encodes information about what happened. Finally, I *understand* that information, that is, I can report that I had a bagel, and that JFK was assassinated, because of the information that this process stored in me, the way it is stored, and my ability to respond to its being there.

When I talk about documented historical events different processes are involved, but the same pattern of reasoning appears: events occur, some people perceive them, and the events are subsequently recorded in various kinds of documents. These documents may persist into the present, or we may only have copies of copies of copies. So long as the processes that produce those copies reliably preserve the information originally contained in the first records, we can reach conclusions about the past by reading these records. The same point applies to geological evidence and the conclusions we draw from it. Processes take place during glaciation, when an organism is fossilized, as sediment is laid down, and so on. These processes leave traces that persist over time. (Note that this persisting is really just another kind of process). As a result, we can find out about glaciation, geology, and past life by studying the traces that remain today. In general, a present *trace* encodes information about a past event because of a *process* that links that event to the trace.

THE SKEPTICAL ARGUMENT

Now we're ready to face a logical problem. All processes take place over time. So any knowledge we have about processes depends on having knowledge about the past. To justify my knowledge claims about the past, I need more than the present traces. I also need to know about the processes that have produced them. This means that I need knowledge of the past to justify knowledge claims about the past. We seem to be stuck in a circle: to justify a claim about the past, we need to already know something about the past. The conclusion is this: *if* our knowledge about the past has to be founded *exclusively* on knowledge about the present, we can't have knowledge of the past at all.

Radical skepticism about the past is just a step away here: no matter what we *now* observe, we cannot infer anything about what has happened before unless we already have some knowledge of processes, that is, some knowledge about the past. From this point of view we can see why nothing we know purely about the present can rule out the infamous five minute hypothesis, which says that the entire world came into existence only five minutes ago. According to this hypothesis, the present, including our so-called memories, historical records, fossils (both in the ground and in the museums), and other traces of the past is just as we usually think it to be, but the past (except for the last five minutes) is just a big illusion.

It's true that we cannot *prove* that this story is fiction using only our present observations. Without claims about past processes to draw on, we can't make any inferences from present evidence to conclusions about the past. So we can't use our present evidence to reach conclusions about the past at all. This shows that the five minute hypothesis is *logically* compatible with our present evidence, including evidence we normally regard as convincing evidence for familiar claims about the past. (In fact, this does more to show how weak logic is than it does to undermine our knowledge of the past.)

The five-minute hypothesis rejects all of our convictions about the processes that produced these traces, and without them our present evidence imposes no constraints at all on what might have happened in the past.

RESPONDING TO SKEPTICS

Circular logic problems like this are familiar puzzles for philosophers. They go back more than 2,000 years, to an argument sometimes called the wheel, or the *problem of the criterion*. The question of the wheel argument is even more general than our question about the past: how can we have knowledge of truth? We take certain claims to be true, and we reject other claims as false. If someone asks us how we decide what claims are true and what claims are false, our answer must show how we separate these truths and falsehoods. That is, we need to explain the *criterion* by which we tell them apart. Suppose that the criterion we come up with really does support the truth-claims we've made. Then our questioner asks how we know our criterion is a good one. This question puts us in an awkward position.

We can say that there is a higher-order criterion (a criterion for good criteria) that our first criterion meets. But then our questioner can ask how we know *that* criterion is a good one, and we're off on a regress we can't end. We can say instead that the criterion is good because it picks out the sentences we know are true and rejects the ones we know are false. But our questioner will point out that we invoked the criterion in the first place to justify our choice of sentences. So it's *circular* for us to appeal to our chosen sentences to justify the criterion. It seems that we cannot possibly defend our claim to know true sentences from false.

But this is just wholesale skepticism, and wholesale skepticism is *very* unattractive. It would be much better to find a response to the skeptic's challenge, one that leaves us with a healthy, common sense account of what we know and what we don't. My suggestion begins with a diagnosis of how we came to be caught in this little trap: if we assume that we're starting from zero, with *no idea at all* of what is true or how to tell truth from falsehood, then we're in trouble. If we really have no idea what's true or how to tell if a sentence is true, then there's nothing for us to go on, and no bootstrap we can use to lift ourselves up to a defensible account. In fact, from this empty point of view we have no idea of what *true* might mean, and no idea of how to find out! But you can't find something out unless you have some idea of how to tell when you've got it right. You can't even begin to look, when you have no idea what you're looking for.

Let's suppose instead that we start with a reasonable list of things that we think are true, and some criteria for truth, that is, some ways of deciding what's true and what isn't that we think are pretty good. Then we've got something we can build on: we can compare what we take to be true and false with the results we get from applying our criteria to decide what's true and what isn't. We can try to improve our criteria by considering the results of applying different criteria and seeing how they fit with what we take to be true. And we can try to improve on what we think is true by applying our criteria and adjusting our beliefs to fit the results of applying them. Sometimes we will change our ideas about the best criteria and methods; sometimes we will change our ideas about what's true. There are no shortcuts, and no simple rules to follow. But over time this process can improve the *fit* between our criteria and our beliefs: we come to describe ourselves, our methods, and their results in ways that support the conclusion that by using these methods we can reliably know these things about our world.

This is exactly how we will approach knowledge of the past. Skepticism is inescapable if we start with no idea of what happened in the past and how processes connect the past with the present. Without the connections that processes provide, nothing we observe can justify conclusions about the past. But in fact we already know quite a bit about the past, and we understand many of the processes that produced features of the world around us. To improve this knowledge, we compare and test these ideas against the traces we find in the world around us. We observe these processes to extend our understanding of them and the traces they leave behind. We use our understanding of physics and chemistry and other sciences to refine our understanding of processes and to identify new processes that might occur. This is how we evaluate and justify our claims about the past. As time goes on, science extends and refines our knowledge of traces and our accounts of the processes that produced them, honing one against the other.

We evaluate claims about processes based on how well they fit with the traces we observe. To be justified, our accounts of the past have to *cohere* with the present evidence: what we observe about present traces and what we say about the processes there are traces of must agree. The traces we find must be the sort of traces that we would expect those processes to produce. Justification also requires practical *vindication*. We have to *apply* our views, using our present ideas about processes and the traces we have already observed, to make predictions about further sorts of traces we can expect to observe. And we have to do this successfully: the observation we go on to make must match our predictions. By these standards our scientific accounts of the past are convincingly supported by the evidence scientists have gathered.

IDLE AND SELECTIVE SKEPTICISM

In the past, many philosophers looked for ways to settle questions with certainty. Today, however, most philosophers have adopted a point of view called *fallibilism*. Fallibilists hold that there is always a possibility of error in our beliefs. Even something we see with our own eyes could turn out to be an illusion. A simple, carefully checked mathematical calculation could turn out to contain a mistake. So fallibilists hold that we *never* have absolute proof of anything, and everything we believe is open to correction. The view of evidence for claims about the past we've just adopted accepts this lesson: the process of testing and refining that we described earlier can always lead to new ideas. Further evidence could come along to change what we now believe about the past.

But we need to distinguish between this possibility and the skeptic's claim that we know nothing at all about the past. The mere possibility that further evidence could overturn a belief doesn't mean that it will. More importantly, it doesn't mean that the evidence we have doesn't *really* support that belief. All it means is that every belief is vulnerable, and just *might* come to be questioned.

This response to skepticism also implies that we can't put all of our beliefs in question at once. Changing our beliefs is like repairing a ship at sea—any part of the ship may need work, but there is no dry-dock where we can dismantle the entire ship and rebuild it from the keel up. We have to work piece by piece, replacing and rebuilding as we go. The wheel argument shows that if we did suddenly suspend all of our commitments, there would be no way back, no way to rebuild our knowledge from scratch.

Idle skeptics respond to fallibilism by giving up. If nothing is ever settled with certainty once and for all, why bother with evidence and reasoning at all? But

David Hume pointed out long ago that idle skepticism is unsustainable. When he left his study and went out into the street, when he had to interact with people and the world, Hume found his skepticism vanished and common sense took over. We can only be idle skeptics during those brief moments when we're actually idle.

On the other hand, *selective* skepticism allows more scope for hanging onto skeptical attitudes. The selective skeptic is someone who only applies skeptical doubts to a particular issue or area of knowledge.

We all have beliefs that matter to us, and we often resist evidence that goes against them. It's very common for religious and moral beliefs to have this status: a serious discussion of such beliefs is always difficult and often uncomfortable. Some people prefer to avoid discussing them altogether. One way to avoid having to examine certain beliefs is to be a selective skeptic, and apply skeptical arguments to any ideas or evidence that puts these beliefs in question. Whatever the evidence, it can be dismissed with idle skeptical objections. Such people often insist that the evidence doesn't *prove* the point in question, ignoring the fact that the evidence they accept in other areas doesn't amount to proof either. They also dismiss the evidence as unreliable, while ignoring how convincingly it fits together and how well tested and supported it really is.

For example, young-earth creationists complain that the processes scientists use to explain present traces aren't completely detailed in every way. They also complain that the traces found so far leave gaps in our evidence. These complaints are *selective*, since the same people do accept evidence in other areas of science like physics and chemistry. But the evidence of lab experiments doesn't conclusively *prove* the conclusions scientists draw about physics or chemistry either. Details are always missing in our theoretical explanations of lab results, and lab measurements always leave gaps in our evidence about the processes under study. Geology and evolution are targets for these people because these sciences directly contradict some of their most-treasured beliefs. But the evidence for these sciences is no weaker than the evidence for other sciences they do accept.

What makes this selective skepticism easier to hold onto than idle skepticism about everything is that we don't really *use* all the beliefs we have. You probably believe that the earth is roughly spherical, that it spins on its axis once a day, and that it orbits the sun once every year. But how often do you actually use these beliefs, say, to explain why the top of a ship's mast is the last thing we can see as it sails over the horizon, or how it is that day and night follow each other in succession, or how the position of the sun against the stars shifts from one season to another? It's hard to be an idle skeptic about everything because we have to act in the world. If I don't know where the fridge is or how to open it, I have no idea where to go or what to do if I want a snack. But it's easy to be skeptical about beliefs that I don't actually use!

YOUNG-EARTH CREATIONISM

Young-earth creationists' approach to geological dating is a good example of this kind of skepticism. Young-earth creationists insist on a literal reading of the book of Genesis, which describes the creation of *everything* as taking place over six days. Following the creation story in Genesis, the Bible provides information on the ages of individuals and their descendants, bringing us up to Noah's flood, and then beyond the flood to independently datable events. Studying these documents carefully and adding up these numbers, a seventeenth-century Irish Archbishop named James Ussher concluded that the world had been created in 4004 B.C.E., on Sunday October 23 (a date that was subsequently inscribed in the margins of some editions of the Bible). This was not the work of some strange eccentric: Ussher was part of a scholarly tradition including famous figures like Isaac Newton and Martin Luther. These scholars didn't all agree on the precise details: there are differences between different biblical texts, and the link between the biblical account and historically known dates is tricky to establish. But a good ball-park figure for the age this method gives is 6,000 years. A 6,000 year history also fits well with a Jewish Talmudic tradition that divides the world's history into three parts of 2,000 years each, with the Messiah scheduled to appear at the 4,000 year mark. Passing lightly over the textual issues, today's young-earth creationists typically focus on limiting the age of the earth (and the rest of the universe, too) to something on the order of 10,000 years.

But scientific measurements put the age of the earth at about 4.6 billion years; the universe as a whole is believed to be about 14 billion years old. The first figure is based on the decay of radioactive atoms in meteorites, which formed during the birth of our solar system when earth and the other planets also formed. Since then, meteorites have traveled through space largely unchanged, which makes them good material for these measurements. The second figure draws on astronomical observations of how far away distant galaxies are and how fast they moving away from us, together with the big bang model of cosmology. The big bang is supported in turn by a wide range of observations, including the afterglow of the big bang, the cosmic microwave background radiation.

These times fit together nicely—the amounts of heavier elements in the sun mark it as a 2nd generation star, based on how these elements are produced in stars. Our understanding of nuclear fusion allows stars like the sun (now middle-aged) a lifespan of about 10 billion years. The rest of our scientific chronology for earth fits beautifully within this framework. We have used radioactive decay to date the oldest rocks found on earth to about 3.8 billion years ago (The earth took as much as 500 million years to develop a solid surface, and most older rocks have been recycled over time.) Fossil evidence of life first appears in the form of bacteria, around 3.5 billion years. Cells like our own, called eukaryotes because most of their DNA is confined inside a nucleus, make their first appearance by about 1 billion years ago. Signs of multi-cellular organisms appear first in trace fossils as old as 1.2 billion years, and later as sparsely fossilized soft-body forms, including the Edicaran fauna dating from up to 600 million years ago. Then about 550 million years ago, near the beginning of the Cambrian period, we find widespread fossils with hard parts including shells, and, shortly afterwards, skeletons like our own.

Geological dating uses many different methods, each with a long history of success at interpreting its special data and avoiding or detecting errors that can arise. Some very simple common-sense methods are surprisingly powerful: dating based on tree rings (dendrochronology) has allowed us to date structures and events some thousands of years past just by counting tree rings. Geologists have also used deep beds showing annual sedimentary layers called *varves* to measure periods of time up to millions of years. Varves form because of seasonal variation in the sediment being laid down; each varve contains a couplet of two distinct layers. The Green River Shales in Colorado, Utah and Wyoming are about 2,500 feet thick; 75 feet of the formation contain about 225,000 couplets. Glaciologists have counted annual layers of snow recorded in the ice of continental glaciers, back as far as several hundred thousand years before the present. They have also used air bubbles trapped in the ice as time capsules providing samples of long-ago atmospheres whose isotopic composition reflects ancient climates. More tentatively, geologists have made estimates of geological dates based on measurements of erosion rates, sedimentation rates, and other long-term, more or less steady processes. Stratigraphy (layering and other relations between bodies of rock) and biostratigraphy (the regular order in which various kinds of fossils appear in layers of sedimentary rock) have been used to reconstruct the sequence of events in the earth's history—a sequence that tests the ages determined by different kinds of geological clocks. Other measurements connect the slow drift of the continents to so-called magnetic arrows in igneous rock, pointing to the position of the magnetic north pole at the time the rocks solidified.

However, the most famous forms of geological dating involve the decay of radioactive elements—a very steady and well-understood physical process. Radiological dating is based on fundamental discoveries about the nature of atoms. Towards the end of the nineteenth century, scientists realized that certain elements were *radioactive*. Over time these elements break down into other elements. Work by Pierre and Marie Curie, Ernest Rutherford and many others established the fundamental characteristics of nuclear decay. Taking place within the atomic nucleus, it is unaffected by the variations in chemistry, temperature, pressure, and other conditions we find on earth. Spontaneous decay proceeds at a steady rate, with half of the radioactive nuclei in a sample decaying in a characteristic time known as the half-life. This rate can be changed for some kinds of radioactive atom, depending on the decay process involved—but only by very small amounts. Exposure to high levels of radiation can also speed up decay. But the levels of radiation required do not occur on or in the earth's crust, a fact for which we must all be grateful! Carbon dating is just one technique of this kind. It involves measuring the amount of radioactive carbon 14 in a sample of material. Living things take up carbon 14 from the air they breathe; carbon 14 levels in the air remain fairly steady because as carbon 14 decays, radiation from the sun creates more to replace it. But once an organism dies, it no longer takes up carbon 14 from the environment, and its carbon 14 level slowly declines. So this technique is only used to date organic remains like wood, plant fibres, or bone. Moreover, because the half-life of carbon 14 is only 5,730 years, we can only use this technique to measure ages up to about 50,000 years. To measure older dates we need to use elements with longer half-lives.

The results of these methods provide a consistent timeline for geological history, reaching back to the earth's formation 4.6 billion years ago. The order independently determined by stratigraphy fits with the times we get from radio-active dating techniques. Our understanding of the decay rates of different iso-topes of uranium, potassium, thorium, and other radioactive materials fits with fundamental physics, and allows us to test and compare results from differ-ent techniques. Knowing the detailed chemistry of various minerals, sampling techniques, and using a clever method called *isochron* dating, scientists also detect and correct for changes in the samples that can otherwise give distorted results. Rocks on different continents, dated both stratigraphically and radio-logically, reveal both the slow shift of the magnetic pole and occasional sudden reversals of the earth's magnetic field. The evidence on different continents fits together only if we allow for the slow drift of the continents themselves. And the motion of the continents is revealed in many other ways as well, including stripes of different magnetic orientation spreading out from mid-ocean ridges where new crust forms, the distribution of life at different times in earth's history, and mountain ranges and other relics of ancient collisions between continents.

The general ideas about testing and supporting claims about the past we discussed above apply beautifully to these methods of geological dating. We have detailed models of the processes; these models pass the test of explain-ing and predicting detailed features of the evidence we apply them to; and the result, taken altogether, is a rich and coherent account of how today's earth came to be.

Contrast this with the views of young-earth creationists. They maintain that the history of the universe reaches back no more than about 10 thousand years. This implies that light from most of the stars in our galaxy (let alone those from other galaxies) has not had time to reach us here on earth. It implies that radiological dating in general, even carbon 14 dating, which only reaches back about 50 thousand years, is uniformly wrong—not just in error, but com-pletely and systematically off-base, despite the rich agreement of different methods based on different data. It implies that the sequence of different forms of life consistently seen in sedimentary rocks worldwide is some kind of illu-sion, not a record of life's history. It implies that volcanoes like Etna formed entirely within the last 4,000 years or so, since Etna arises through sediments that young earth creationists hold were laid down in the flood. And yet there are no historical records of the devastatingly frequent and violent eruptions that would be required to accomplish this. It implies that soft sediments can harden enough to support hundreds of feet of material sitting above them in as little as a year. (They must make this absurd claim in order to explain the Grand Canyon and many other such dramatic features.) It implies that a single

Isochron Dating

Given a sample of rock formed from the same source, we can measure its age by the isochron method. Rocks contain different minerals. The key to isochron dating is that each of these different minerals forms by drawing the necessary atoms of each of its elements from the same pool of materials. Different isotopes of the same element are chemically indistinguishable. So the isotopes of each element in each mineral start out at the same ratio found in that initial pool. However, some of the elements have radioactive forms; these are called parent isotopes. Radioactive atoms decay over time into daughter atoms, which are isotopes of another element. The more parent atoms originally present, of course, the greater the number of daughter atoms produced over time. These radiogenic daughter atoms are a particular isotope of the daughter element. As they accumulate in a sample of mineral, the ratio of daughter isotopes in the mineral changes.

Let's call the parent element P and the radioactive isotope of P, Pr. We'll call the daughter element D and label its radiogenic isotope Dr, while we call a non-radiogenic isotope of the same element Dn. As Pr decays over time, Dr builds up in the rock. This is the basis of all radiological dating techniques. But isochron dating makes a particularly subtle use of the data we can gather to guard against problems that other radiological dating techniques can encounter.

A mineral that initially took up none of the P will contain no Pr, and won't produce any more Dr over time; the amount of Dr present in the mineral will be just what the mineral took up initially from the pool. In such a mineral the ratio of different isotopes of the daughter element will not change over time. But a mineral that contains a lot of the parent element will produce a lot of the daughter isotope, changing the ratio of isotopes of the daughter element. Different minerals will take up different amounts of P, including proportionate amounts of Pr. So in each mineral within the rock the ratio of the daughter isotopes will change over time at a rate proportional to the amount of P present in the mineral.

Now we can plot measurements of the ratio Dr/Dn against measurements of the ratio of Pr/Dn present in each sample. The result should be a straight line. The line is flat when the rock has just formed because at the beginning the Dr/Dn ratio in every mineral is just the original ratio of Dr/Dn in the pool of material the rock formed from. But as time passes, minerals with more P (and Pr) produce proportionately more Dr. The line stays straight, but its slope increases as time passes. So the slope of this *isochron line* is a measure of the time since the rock formed.

As atoms of Pr decay into Dr, the ratio Pr/Dn changes, subtracting from the numerator (Pr). At the same time, each decay adds to the numerator of Dr/Dn. So each of these lines joining earlier to later samples of our minerals has a slope of -1, and the distance each mineral travels along these lines over the time from 0 to x is proportional to the amount of Pr in the mineral at a given time.

This technique is not foolproof—if the rock is heated enough for the minerals to reform, the clock can be reset to a new zero time. If it is heated just enough to partly reset the clock, things can get even more difficult to sort out. But such episodes of heating can often be detected or ruled out.

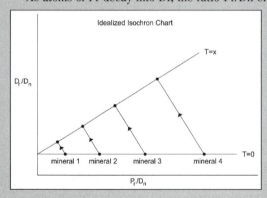

An Idealized Isochron Chart. Dn is a non-radiogenic isotope of the daughter element, Pr is the radioactive isotope of the parent element and Dr is the radiogenic isotope of the daughter element. Illustration by Jeff Dixon.

Moreover, isochron dating does provide a direct check on whether losses of daughter isotopes or other kinds of interference have occurred, since such losses affect different minerals differently, destroying the straight line. The fit of our data points to a straight line thus becomes a measure of how reliable our age estimate is. A further advantage is that isochron dating adjusts automatically for whatever amount of Dr was initially present in the pool.

world wide flood is responsible for most of the geological record, even though so many details of that record indicate long, slow periods of erosion and sedimentation. To choose just one example, consider the varves described above. How could such delicate, regular structures of thin couplets (pairs of layers), just like those we see being laid down in lake beds today, be the product of a single, violent flood lasting only about a year? Young-earth creationism implies that life on earth today spread from a single site (Noah's landfall). But different forms of life are distributed across the globe today in ways that reflect the fossil record of each individual place, and the distribution of fossil forms reflects the known history of continental drift. It implies that all the animals around us descended from single pairs (with the exception of Noah's so-called clean animals, of which it seems seven pairs were saved). But the genetic diversity we now see in animals is so wide that these creationists have to postulate far higher rates of mutation and evolution than evolutionary theorists accept. It implies that all the earth's languages originated recently, at a single place and time. But the distribution of languages across the world today and the connections amongst them show that today's languages are descended from older forms, and spread across the world with human migration in patterns that just don't fit with the flood myth.

Young-earth creationism is an astonishing exercise in selective skepticism. It demands that we ignore all this evidence and much, much more. It is true that traces left today are never enough to completely reconstruct the past: even the details of recent historical events like John F. Kennedy's assassination are disputed, despite the fact that we have film of the event. But the evidence we have is more than enough to reveal both the outlines and many details: the fact that Kennedy *was* assassinated is settled (despite tabloid rumors). And the fact that life has evolved is also settled. Scientists have been finding, extending, and refining evidence supporting this conclusion over the last four hundred years.

The last serious naturalist to defend young-earth creationism was Philip Henry Gosse, a contemporary of Charles Darwin's. As a biblical literalist, he could not accept either the great age of the earth or the history of life that geology and paleontology require. Nevertheless, as a professional naturalist Gosse frankly acknowledged what today's young-earth creationists will not: that the evidence was against him. So he advocated a radical skepticism about the distant past, claiming that the earth had been created with a so-called false past built into it. Still, Gosse believed that science could legitimately study this false past *as if* it were real: even though it never really happened, it still reflected a coherent aspect of the creation. His book was titled *Omphalos*, which is the Greek word for navel (a reference to the long theological debate over whether Adam, who had never been in a mother's womb or had an umbilical cord, would have had a navel—of course this theological debate has never been satisfactorily resolved). Gosse, of course, sided with those who believed Adam did have a navel, as a part of the false past built into creation. Gosse hoped that his maneuver would reconcile religion with science, but to his great disappointment he was attacked by both sides. Scientists pointed out that there was no scientifically motivated line to draw between the past that Gosse accepted as real and the past he wanted to treat as illusory. And Gosse's religious critics rejected the notion that God would build deceptive evidence of a false past into the created earth.

The scientific evidence has gotten stronger, more detailed, and much more complete since Gosse's time. No one who takes science seriously today defends young-earth creationism. Only a radical skeptic can manage the trick—but radical skepticism is so corrosive that it destroys both science and common-sense together. Today's young-earth creationists spend most of their time taking skeptical potshots at specific details of the scientific evidence. They spend the rest of their time carefully avoiding Gosse's conclusion: only radical skepticism about the past can rescue a literal reading of Genesis. This is the last refuge of those who reject what scientific evidence shows. But, as the fourth-century Christian thinker Augustine pointed out long ago, when the religious talk nonsense about scientific matters, it only exposes religion to scorn:

> Often a non-Christian knows something about the earth, the heavens, and the other parts of the world, about the motions and orbits of the stars and even their sizes and distances, ... and this knowledge he holds with certainty from reason and experience. It is thus offensive and disgraceful for an unbeliever to hear a Christian talk nonsense about such things, claiming that what he is saying is based in Scripture. We should do all that we can to avoid such an embarrassing situation, which people see as ignorance in the Christian and laugh to scorn.
>
> *De Genesi ad litteram libri duodecim*
> (The Literal Meaning of Genesis)

As I said at the outset, the main aim of this book is to present a history of how the case for evolution developed over time. Evidence for evolution—both in support of the descent of all life on earth from a common ancestor, and in support of Darwin's mechanism of natural selection, did not appear suddenly, in its final form, in *On the Origin of Species*. Some of the evidence predates Darwin, and much more has been discovered since Darwin first presented his ideas. New details and insights continue to emerge today, in the work of biologists, biochemists, taxonomists, geneticists, and paleontologists around the world.

1

LIFE'S HISTORY

LIVING KINDS

The natural world is full of living things, both obvious and hidden. There are three-toed sloths and *E. coli* bacteria, silver maples and monarch butterflies, lamprey eels and oyster mushrooms, paramecia and goose barnacles. These living things come divided up into different *kinds,* and when each kind reproduces, the offspring belong to the same kind as the parent organisms. These are pretty straightforward observations, although we had to wait for the invention of the microscope to find out that there were many living things too small to see with the naked eye.

Observations of life today are limited by our point of view in time. Our astronomical observations were once similarly limited by our point of view in space: up until the last 50 years, we could only see the sky from the surface of our spinning, orbiting planet. As a result, it took a long time, lots of observations and much careful calculation before we finally realized that it was our point of view and not the rest of the universe that was spinning around every 24 hours. Similarly, when we look at living things, we see life only as it is now. It was not until the late eighteenth and early nineteenth centuries that the outline of the earth's history began to become clear to scientists. One important lesson that they learned then was that life on earth has changed over time.

Finding out that the earth moves and is not the center of the universe transformed our understanding of the universe and our place in it. The discovery that life on earth has a history, and especially the recognition of our own place as newcomers to this world, has had a similar impact.

The Greek philosopher and scientist Aristotle was an important early thinker in biology. He drew a very general distinction between the matter that things are made of and the form that matter has. According to Aristotle, it's the form that makes some piece of matter into a particular kind of thing. He used this connection between form and kinds of thing to systematize his ideas about life: living things come in kinds because the members of a kind share a *form* that all living things of that kind possess—a form that parents pass on to their offspring.

Aristotle divided the basic faculties (abilities) of living things into three categories: nutrition, motion, and thought. He held that all living things possessed what he called a nutritive soul, or in other words, a faculty of nourishing themselves and growing. But, unlike plants, animals also have a sensitive or appetitive soul, which enables them to perceive, to *want* or have desires, and to move around. Finally, unlike other animals, humans possess a rational soul, which enables us to think and reason. Living things are matter *informed* by one or more such souls.

Figure 1.1: Aristotle was an amazingly energetic and wide-ranging scholar. Trained in Plato's academy in Athens, he served as tutor to Alexander the Great. He wrote on almost every branch of learning, from the sciences—physics, astronomy and biology, to metaphysics, logic, ethics, and more. As much as 80% of his work has been lost, but what remains, gathered by Arab scholars and transmitted to Europe in the twelfth and thirteenth centuries, is still a rich legacy. European science, philosophy and theology were transformed by this late encounter with Aristotle's works.

For Aristotle each living thing is a *substance,* combining matter and form. Two members of the same species have the same form—but they are distinct individuals because the matter is different in each. Interaction of the particular matter with the shared form explains both the general resemblance of members of the same species and the small differences between individuals even when they belong to the same species.

Aristotle's species were at least potentially eternal: parents passed the form, the *essence* of their kind, on to their offspring unaltered. So for Aristotle, as long as the line of descent continues unbroken, the very same species with the very same form continues. A related point is that according to Aristotle, new species could not arise. His account of life didn't allow organisms to exchange old essences for new; neither did he allow new living kinds to arise spontaneously. So for Aristotle there could be no real history of life—individuals will come and go, but the kinds would stay the same forever, although some might become extinct and disappear.

Another important observation about life is that the kinds of living things can be grouped in kinds of their own. Although they are clearly different species, wolves, coyotes, domesticated dogs, and the hunting dogs of Africa are very similar in many ways. Another familiar group of carnivores, including lions and tigers, ocelots and leopards, jaguars and serval cats, shares many cat features, though they differ widely in size, color, social behavior, hunting techniques, and more. Plants no less than animals come both in specific kinds and in kinds of kinds, distinguished by types of leaf, early growth patterns, the details of how they reproduce, and so on. At a still more general level, mammals, which nurse their young, and have teeth with different shapes, warm blood, fur or hair, and seven neck vertebrae, all resemble each other in these and other ways.

These patterns of resemblance were the key to the Carolus Linnaeus's (1707–1778) great book, *Systema Naturae*. Published in 1735, Linnaeus's book laid the foundations of biological taxonomy. A *taxonomy* is a system of classification, that is, a regular way of grouping things into kinds. Linnaeus's system was *hierarchical:* it included different levels of kinds, beginning with basic kinds and then grouping these in turn into kinds at higher and higher levels. In Linnaeus's system the first level groups individuals into species. The second groups species together into genera, and so on up the hierarchy. Linnaeus continued to refine his system for many years. In 1755, he published *Species Plantarum*, coining the oldest scientific names for plants that are still in use today. The tenth edition of *Systema Naturae*, which contains the oldest scientific names for animals still in use, was published in 1758.

Linnaeus's taxonomy assigns each individual living thing to a species. Every species, in turn, is assigned to a *genus* (Latin for kind). The species are identified by their *binomial* names, which combine the generic name (capitalized) with the name of the particular species. For example, human beings are classified as *Homo sapiens*, where *Homo* is the human genus, and *sapiens* is our particular species. Similarly, the house cat is *Felis domesticus*, the mouse is *Mus musculus*, and one kind of fruit fly, famous for its role in genetic research, is *Drosphila melanogaster*. This system of names allows scientists all over the world to identify the kinds of organisms others are talking about in a consistent and reliable way.

Above the lowest two levels of species and genus, Linnaeus proposed higher-level categories called *ranks: orders* group similar genera (the plural of genus) together, *classes* group orders, and *kingdoms* group classes. Today taxonomists also recognize at least three further levels of ranks: *families, phyla* and *superkingdoms*. Families group similar genera together, and are themselves grouped into orders (or sometimes into *superfamilies*, which are then gathered into orders). Phyla (singular phylum) group related classes together. Finally, *superkingdoms* mark the most fundamental division between three basic kinds of life: archaebacteria, bacteria, and eukaryotes.

At first Linnaeus followed Aristotle, thinking of each species as something real and unchangeable. Later, after studying how different looking plants could be produced by hybridizing different species and noticing that some plants seemed to change their characteristics when raised in a new climate, Linnaeus suggested that many of today's species might have arisen when two species combined to form a new, hybrid species. He even came to suspect that new genera might be formed in this way. But for Linnaeus these new developments were already implicit in the originally created species; the notion that such changes might continue indefinitely awaited new evidence, and new ideas.

ANCIENT LIFE

§26 In very ancient times the nearby seas contained animals and molluscs that are not found there now.

G. W. Leibniz, *Protogaea.* (1691, §26)

The idea that life has a history, that it has changed dramatically over time, was once a new and radical thought. The study of ancient life is called *paleontology*. Its modern form began to emerge in the sixteenth and seventeenth centuries, alongside its sister science of geology. The argument that persuaded those scientists that life had changed is very simple. It is based on two conclusions that they reached after great effort and much dispute; the evidence gathered since that time has only added to the support for both of them:

1. Fossils are relics of ancient life.
2. Layers of sedimentary rock were laid down in sequence, with the oldest layers lowest and younger layers higher up.

The things we call fossils[1] have been familiar to observers of nature since ancient times. Though they look just like parts of living things, scholars haven't always believed that these so-called figured stones really are remains or traces of ancient life. Many authorities in medieval times held that fossils were produced by natural forces. These forces were supposedly related to the forces that produce living things. Acting on different kinds of materials, they produced shapes in rock that resemble shells, bones, teeth, and so on. A stone-making force (*vis lapidifica*) was proposed by Avicenna; a playful, plastic (i.e., shape-forming) force (*virtus formativa*) by Albertus Magnus. A seminal aura (a seed or fertility principle), and the power of spontaneous generation (still widely believed to occur at that time) acting on inorganic matter, were also invoked to explain fossils. These views provide an easy answer to the puzzle of the shellfish in the mountain: there never was a shellfish there at all—what we see are just strange facsimiles of shellfish, formed out of stone by these forces of nature.

As astronomy was moving from the old earth-centered view of the solar system to the sun-centered view proposed by Copernicus and defended by Galileo (who was branded a heretic and imprisoned by the church), debates between scholars who defended this position and scholars who believed fossils really were the remains of living thing were underway too. One of the latter group, Leonardo da Vinci wrote a lengthy treatise on water in which he addressed the puzzle of how fossil shells come to be found so far from the sea, even high in the mountains (Cutler 2003, 46–47). Leonardo argued that a popular explanation of his day, that the shellfish had migrated or been swept up to the mountain during Noah's flood, couldn't hold water: shellfish don't move fast enough to get from the beaches to the mountain tops in time. Moreover, floods transport material downhill, not up. Leonardo also pointed out that fossils are found *inside* rocks, not just on the surface, something that a violent flood carrying them up to the mountaintop couldn't explain.

Leonardo went beyond these negative points to argue that the fossils really were the remains of ancient shellfish. The details of the fossil shells and their surroundings fit beautifully with Leonardo's position. In some places in the mountains, worm tracks could be seen on the rocky surface. The fossil shells showed yearly growth bands. While some shells were whole, others were marked by holes or broken up: in every way, these fossil shells looked exactly like the shells we find strewn on the seashore today. If they'd been swept up to the mountain tops in a violent flood, the remains wouldn't be spread out as they are, like shells on an ordinary beach today. And if they formed spontaneously in the rock, we have to wonder why there should be broken ones mixed in with the whole. It's hard to imagine any process that would produce all these familiar features other than the ordinary ones that go on today, as shellfish grow, spread, die, and are buried in the mud. Finally, Leonardo also noted that there were several distinct layers of shell-bearing rock in the mountains he explored. A single episode like the biblical flood could not account for this either.

In 1666, a Dane named Neils Stenson, better known today as Nicolaus Steno, joined the court of the Grand Duke Ferdinando II in Florence, Italy. Ferdinando was one of the famous Medici rulers of Florence. His new courtier Steno had been trained as a physician. Despite his youth, he already had a reputation as an expert anatomist. In Florence, he joined the *Academia del Cimento*, the first scientific society dedicated to doing experimental science. Sponsored by Duke Ferdinando and his brother Leopold, the society's members included two of Galileo's students.

During the summer that Steno arrived, a member of the *Academia* named Francesco Redi was performing one of the first controlled (and what must have been one of the worst smelling) experiments in history: he was observing jars of rotten meat and dung from many different animals. The meat and dung were treated in various ways. Then some samples were sealed, some covered with light veils to keep flies off and others were left uncovered. Redi's aim was to see if flies and other vermin would be produced spontaneously. Since nothing but maggots and flies appeared, and they appeared only in the uncovered meat and dung, Redi concluded that the flies that emerge from rotten matter are not spontaneously produced through decay, but instead grown from eggs laid by other flies (Cutler 2003, 48f).

Steno himself was soon at work on a project with a powerful odor of its own. A large great white shark had been caught by local fishermen, who presented to the Duke. The head of the shark was removed, and Steno was called upon to dissect it. Steno prepared a short report on his observations, under the title "On the head of a shark dissected." In his report, he noted something striking about the shark's teeth: they looked exactly like *tonguestones* (*glossopetrae*,

Figure 1.2: A modern great white shark tooth. Image used with the kind permission of JTs Shark's Teeth <http://jtssharksteeth.com/>.

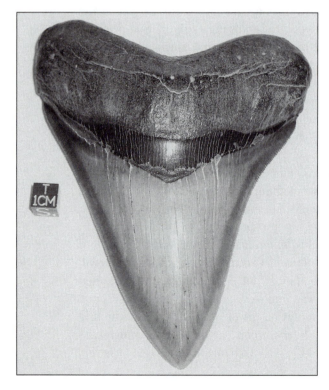

in Latin). These strangely shaped stones, widely believed to have magical powers, were well-known in rocks on the island of Malta. Steno found the similarity so perfect that he became convinced that these strange stones had originally been teeth in the heads of ancient sharks.

To defend this bold conclusion, Steno began a general work on solids within solids, examining different processes by which solids could come to be inside other solids, and describing the clues we can use to decide which of two such solids was solid first.

SEDIMENTS AND STRATIGRAPHY

The problem is a neat one. We see a solid, surrounded on all sides by another solid. As they are now, the first couldn't be placed inside the second without cracking the second open and carving out a space inside to match the shape of the first. Steno concluded that one solid must have formed before the other, and imposed its shape on the second as it hardened in turn.

The principles and clues Steno used to figure out the process are worth thinking through here. First, Steno declared that if one solid's shape is impressed on the other, then the object that did

Figure 1.3: A fossil shark tooth. Note the similarities, both in overall shape and in detail, with the modern tooth in figure 1.2. Image used with the kind permission of Arizona Skies Meteorites <http://www.arizonaskiesmeteorites.com/>. © Arizona Skies Meteorites 2007.

the impressing was solid first. Of course we can't apply this criterion unless we have some notion of what sorts of solid shapes form spontaneously, and what solid materials do or don't generally take on such a shape. But when it comes to fossils, experience answers our question: we see that the shapes of shells and teeth and other parts of living things form spontaneously in plants and animals. On the other hand, sandstone, limestone, and other sedimentary rocks are bulk materials with no characteristic shapes. Even when they contain fossils, the fossils often differ from one formation to another. Since these fossils resemble the shapes of solids that grow spontaneously as parts of living things, it's far more *economical* to suppose that these shapes formed first, in the usual way, that is, as parts of living things. The alternative order of solidification requires that some force produced corresponding hollows that were then were filled in like molds. But this would require brand new processes to form these so-called negative shapes.

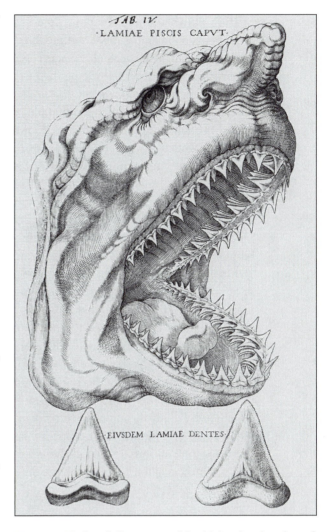

Figure 1.4: Medieval illustration of shark's head and teeth used by Steno. Image courtesy History of Science Collections, University of Oklahoma Libraries; © the Board of Regents of the University of Oklahoma.

Steno also noted that, unlike tree roots growing in soft earth and in stony ground, the shapes of fossils aren't altered by the kind of rock or ground they are found in. The teeth retain their shapes, whether the rock they're in is hard or soft, cracked or solid. Steno concluded that the teeth were solid first, at a time when the rock that solidly surrounds them now was still a fluid sediment. Sediment was laid down around and on top of the teeth, molding itself to their shape. Later, it hardened into rock and was either lifted above the water, or exposed as the waters withdrew.

Steno knew that tonguestone fossils were often made of different material than the teeth they resembled so perfectly. But he had an explanation for this as well—the material of the original tooth had been replaced over time. As fluids moved slowly through the rock, they had dissolved the original tooth material, leaving different minerals behind in its place.

By thinking carefully about the process of sedimentation, Steno reached some important conclusions about sedimentary rock, too. These insights led him to recognize that the history of the earth was written in the rocks, and could be reconstructed by carefully tracing the process back, step by step, to earlier situations. First, Steno noticed that the fossil-containing rocks he examined had a layered structure. Steno claimed that the layers had been deposited as sediment, one after another, and concluded that lower layers were older and layers higher up in the sequence were younger (So long as nothing has turned the whole mass upside down!) This is Steno's *principle of superposition.*

Since the sediments start out as a fluid, Steno concluded that they would spread out to cover the surface they were deposited on. This implied that, where a valley runs between matching beds of sedimentary rock, the sedimentary rock was originally a continuous mass which the formation of the valley had later divided. This is called the principle of *original continuity.*

Finally, Steno concluded that fluid sediments could only be laid down more or less horizontally. This implies that when we find layers of sedimentary rocks that are tilted, the tilting must have happened after the sediments had hardened into rock. This is called the principle of *original horizontality.*

Applying these principles to the geography of Tuscany, Steno constructed the first geological history. His history of Tuscany's landscape includes several stages, beginning with the laying down of the first (lowest) sedimentary rocks in the region. It then proceeds through periods of erosion and collapse, new sedimentary deposits, and more erosion and collapse, to finish with the present structure of the regions' bedrock.

Steno wrote an initial study of these issues, known as Steno's *Prodromus.*[2] He also planned a much larger work on the subject. Sadly, it was never finished,

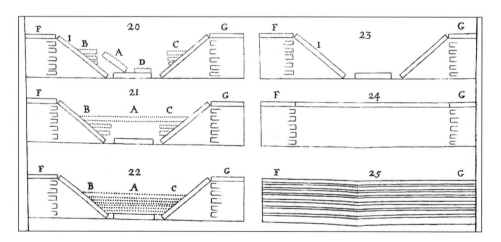

Figure 1.5: Steno's reconstruction of the geological history of Tuscany: beginning with simple landscape of sedimentary layers, erosion, collapse, and the laying down of new sedimentary rocks combine to produce the present landscape. Image courtesy History of Science Collections, University of Oklahoma Libraries; © the Board of Regents of the University of Oklahoma.

and the incomplete manuscript was lost. Steno converted to Catholicism in 1667. Shortly afterwards, at the request of the King of Denmark, he left Italy for Denmark, but after a brief time in Denmark he returned to Italy. Back in Florence he served as tutor to the Duke's son before deciding to become a priest. Steno rose quickly in the Church hierarchy, becoming a Bishop in northern Germany. There he took a strict vow of poverty, became increasingly austere and intensely religious in outlook. Neglecting his health, he died in 1686 at the age of 48.

Some of Steno's contemporaries agreed with his views on fossils and stratigraphy, including important figures like Robert Hooke, John Ray, and G. W. Leibniz. However, it still took nearly one hundred years for these ideas to be generally accepted. By the middle of the eighteenth century it was clear that the earth had a long and complex history. By the late eighteenth century, the history of the earth and the history of life revealed in the fossil record were becoming closely linked.

This development brought an important fact to light: many kinds of living things have become extinct over time. The great paleontologist Georges Cuvier is generally accepted to have been the first to demonstrate this conclusively, when he showed that fossil elephants found in Europe were distinct species from the African and Indian elephants of today. Some scientists, including Thomas Jefferson, resisted this conclusion. Jefferson even funded scientific expeditions into the interior of North America, hoping to find living Mastodons, along with other fossil species he had identified. Extinction was especially problematic for those who believed in the chain of being, a doctrine holding that all the possible grades of existence had been created, and that their continued existence was required for the completeness and perfection of God's creation. But the evidence continued to accumulate. As the fossil record and the earth itself were examined more and more thoroughly, it became obvious: many of the plants and animals that had been found in the fossil record had since died out.

THE PROBLEM OF CORRELATION

By the eighteenth century the basic principles of stratigraphy were well established. Superposition, tilted strata, crosscutting relations, and other clues were used as scientists attempted to unravel the sequence of events revealed by local geology. However, this kind of stratigraphy is strictly local. Only locally can we trace the relations between formations, observe their order of superposition, identify crosscuttings of earlier structures by later ones, and so on. These local methods cannot settle whether an outcropping of limestone in England is older or younger than a sandstone formation in Nova Scotia: neither formation lies above the other, and we cannot trace their relations to other formations across that distance. This is the problem of *correlation*. Local stratigraphy can give us a well-ordered history for each region, but we don't know how the order of events in one place relates to the order of events somewhere else. Is there any way for stratigraphers to correlate times across distant regions?

An eighteenth century figure named Abraham Gotlob Werner proposed one answer. A professor at the mining school in Freiberg, Werner developed both a detailed system for identifying minerals and an ambitious theory of the history of the earth. According to Werner's Neptunism the entire earth was once covered by a deep ocean whose waters were nearly saturated with dissolved minerals. As Descartes suggested long before, Werner held that the waters of the ocean were gradually lost to space. The waters left behind slowly became supersaturated with minerals, and the oldest rock to be found, lowest in the stratigraphic column *all over the world,* was formed by precipitation from this concentrated solution. As the universal ocean gradually receded, later transitional layers combined rock formed by precipitation with rock formed by erosion and sedimentation. Still more recent layers were dominated by sedimentary rocks, including sandstone, limestone, and coal, with an interruption by a return of the sea. Thus Werner held that we could tell when a particular formation was laid down by examining the type of rock it was made of and its place in the local stratigraphy.

Werner's bold effort to establish a universal stratigraphy soon collapsed. Work by Desmarest and others in central France showed that basalt, a rock Werner had declared to be precipitated from the universal ocean, was in fact produced by volcanoes. Werner also held that the heat of subterranean coal fires was responsible for volcanic activity. But these volcanoes rose up through granite and basalt rock that Werner's theory declared to be primitive. Therefore Werner's theory implied that later rocks, including coal, could not lie beneath them. So Werner's stratigraphy couldn't explain what had fueled these volcanoes. In Scotland, James Hutton showed that granite, another of Werner's so-called primitive rocks, was also volcanic. Werner's universal order was coming apart at the seams.

Hutton proposed an alternative approach to geology. On his account, the earth had undergone repeated cycles of erosion, sedimentation, and rock formation, followed by violent collapse and uplift. These cycles had gone on since indefinitely far into the past, and would continue, as far as he could tell, indefinitely far into the future. Hutton concluded that in geology "we see no vestige of a beginning, no prospect of an end." (1788, 304)

Some striking observations on *angular unconformities* contributed to Hutton's vision of deep geological time. An angular unconformity occurs where one sedimentary formation rests on top of the eroded surface of an older formation at a different angle. Thinking through the steps involved in producing them, Hutton saw that angular unconformities provided evidence for at least four cycles of erosion, deposition, subsidence, and uplift. In the first, the lower formation accumulated slowly at the bottom of the sea. In the second, it was uplifted and eroded over another long period of time. In the third, it subsided below the sea again, and the upper formation was slowly deposited on its eroded surface. Finally, in the fourth, the two were uplifted together, and eroded to their present condition. Each of these cycles was of unknown, but clearly vast duration.

At the close of the eighteenth century the vast abyss of past time was becoming increasingly evident. These ideas in turn triggered a dramatic transformation of how scientists conceived our place in the world, extending the change from traditional Ptolemaic earth-centered astronomy to Copernicus's sun-centered model of the solar system. The astronomical shift had taken us from a cosmology in which earth occupied the center of a fairly small universe, to one that placed the sun at the center—but only of the solar system, which itself was located nowhere special, in an immense universe full of other suns. The geological shift reduced recorded human history to a tiny endnote in the history of the earth. Together, they presented a much more modest perspective on our place in the natural world, shifting us from the center-stage of space and time to latecomers on a small and peripheral planet.

As a Plutonist, Hutton held that the driving force behind the uplifts that produced new continents was the heat energy of the deep earth. Assuming that this energy was inexhaustible (at least as far as could be seen), Hutton's geology was an endless cycle of processes, rather than an arrow of change pointing from a previous state to the present and beyond (Gould, 1987).

Whether Hutton was right or not about the earth's ancient past, it was clear that Werner's vision of stratigraphic correlation based on a universal ordering of rock types was untenable. Whatever his success at interpreting the local

Figure 1.6: The Hutton Unconformity. Jedburgh, Scotland (John Clerk, 1787). Image courtesy History of Science Collections, University of Oklahoma Libraries; © the Board of Regents of the University of Oklahoma.

stratigraphy of Saxony, Werner's so-called universal sequence of rock types didn't apply everywhere. The problem of correlating the stratigraphy of distant locations remained unsolved.

BIOSTRATIGRAPHY

By the beginning of the nineteenth century, a very different clue to the ordering of formations around the earth was coming to be recognized: it was based on the growing evidence that life itself had changed through time. Wherever they were found, certain types of fossils appeared in the same order in the local stratigraphy. Better still, some could be found in far-apart locations, always confined to particular layers of rock. These index fossils allowed geologists to correlate the formations they studied in a regular, worldwide order.

Let's pause here to think this method through. A simple analogy helps to illustrate the reasoning. I keep a pretty messy desk—there are always different projects that I'm working on, and I keep papers related to each in more or less separate piles. For the most part, the piles are stable for a long while, with additions made to the tops of the piles as new material for that project arrives on my desk. So Steno's principle of superposition applies: papers on the top are generally more recent than papers lower down in the pile. And since I usually put papers down face up, I can even tell when a particular pile has been turned over.

Figure 1.7: A typical state of a messy desk. Image used with the kind permission of Matthew Hodgson, and TheOneRing.net.

Now suppose you wanted to know, not just the order of arrival for particular papers in a given pile, but the order of arrival for papers in different piles. What we've established so far isn't enough. On one hand, I might always work on my projects one at a time, so that the pile over on the left was completed before the pile on the right was even begun. On the other hand, I might work on them in parallel, so that the two piles began at the same time, and each has grown, step by step, alongside the other. We need some more information before we can decide between these hypotheses.

One possibility, which leads immediately to a promising and testable proposal, is that you will find marks on the papers that obey a simple pattern. If there are marks that many of the papers bear, and that always occur in a certain order in each pile, then you might hypothesize that they *date* the papers that bear the marks. Now, it doesn't really matter what the marks look like. You don't have to be able to read them. You don't even have to suppose they are part of some kind of language. And some marks may not appear at all in some piles (after all, I don't have to work on every project every day). All that matters is that you can tell one type of mark from another, and that in every pile where two marks appear, they appear in the same order.

You can test your hypothesis by examining new piles. Of course, in the piles that seem to have been turned over, your hypothesis predicts that the usual order of the marks will be reversed. This helps rule out the possibility that the depth of the pile, not the time at which the paper was placed in the pile, is what settles which marks appear on which papers. (You might also rule this possibility out by making observations on piles of varying depths).

If I sometimes shove part of one pile over on top of another, the usual order will apply within the original pile and within the part shoved on top. But there could be a break in the order where the bottom pile stops and the added pile begins. So variations in the order can result when such processes act on the piles. To help verify the idea that this has happened in a particular case, you could look for clues, like creases and folds in the papers around the boundary between the original piles, indicating that such shovings have happened.

The order of different fossils and groups of fossils in the stratigraphic record follows exactly this kind of pattern. In each locale, when we examine stratigraphic relations between different formations the characteristic fossils appearing in those formations show a regular order. This is the basis for the names geologists gave to the three main divisions of the *Phanerozoic Eon* (the time during which macroscopic fossils are common in sedimentary rocks): the Paleozoic Era means the era of *ancient life*, the Mesozoic Era the era of *middle life* and the Cenozoic Era the era of *recent life*. These names were coined long before evolution was accepted. The stratigraphic evidence showed that in the distant past, living things were very different from today's. It also showed that over time life has become more and more like the familiar life we see around us today.

An English surveyor and engineer named William Smith was one of the first to notice the regular appearance of fossils in particular strata. He impressed other fossil collectors with his ability to tell them what rock formations various fossils in their collections had come from. Studying the formations revealed in mines, canal excavations, and other engineering works as well as natural out-croppings, Smith had recognized that it was very difficult to tell one sandstone formation from another by looking at the rock. But fossils were a give-away. Drawing on these sequences of fossils, Smith worked for years to prepare the first geological map of England. Smith saw clearly just how valuable such a map could be: it would reveal places where coal and many other resources might be found, and (even more helpfully) it would also reveal places where digging for them would be pointless.

The job of systematically classifying and correlating geological periods around the world was well underway in the early nineteenth century. The work combined Steno's principles with growing knowledge of the particular fossils appearing in different periods. William Smith continued to prepare his geologi-cal map of England. Together with Alphonse Brogniart, the great paleontologist and anatomist George Cuvier began to establish a systematic ordering of the Cenozoic fossils appearing in the Paris Basin. What they observed there was a sequence of fresh-water sedimentary layers, separated by ocean sediments. The fossils they found changed from one fresh-water layer to the next. Cuvier and Brogniart concluded that the ocean sediments represented catastrophic incur-sions of the sea onto the land, which had wiped out all the plants and animals in the area. After the sea retreated, new forms of life migrated in to replace them, only to be wiped in the next catastrophe.

The early Paleozoic was first divided in the 1830s, between Sedgwick's Cambrian and Murchison's Silurian. When these were found to overlap, a long and unfriendly dispute arose over which system would have to shrink. The issue was finally settled by Lapworth in 1879, who identified three distinct fossil faunas in the early Paleozoic, and proposed a division of the Paleozoic on these fossils. Lapworth's work was based especially on fossils of graptolites—an ancient *hemichordate,* related both to starfish and to true chordates like ourselves. Graptolites were common in these formations, and Lapworth was an expert on them. Lapworth's solution was to insert a new period, named the Ordovician, between the Cambrian and the Silurian. The later Paleozoic was divided into the Devonian, Carboniferous, and Permian periods.

At the start of the Paleozoic era all life lived exclusively in water. The first indications of life on land are tracks dating to the Ordovician, but these are rare until the Silurian and only show up in sea-side locations. Plant fossils first appear on land in the late Silurian and early Devonian. By the end of the Paleozoic, active land-based animals included mammal-like rep-tiles and the ancestors of dinosaurs and other *archosaurs* (ruling reptiles). Large trees and a wide variety of other land plants had also appeared. Still, Paleozoic landscapes would look very strange to us today: both animals and plants were very different from those we see today. In particular, the flowering plants that dominate the land now did not arise until late in the Mesozoic.

Nineteenth century biostratigraphers divided the *phanerozoic* into its three main parts, the Paleozoic, Mesozoic, and Cenozoic eras. Each of these was in turn divided into periods: the Paleozoic era includes the Cambrian, Ordovician, Silurian, Devonian, Carboniferous, and Permian periods; the Mesozoic begins with the Triassic, followed by the Jurassic and, last, the Cretaceous periods; finally, the Cenozoic includes the Tertiary period (made up of the Paleocene, Eocene, Oligocene, Miocene, and Pliocene epochs), and the Quaternary period (made up of the Pleistocene and the Holocene epochs). Geologists distinguish the rocks we now observe, divided up into *systems,* from the times when they were formed. But the same names are used for both. Each of

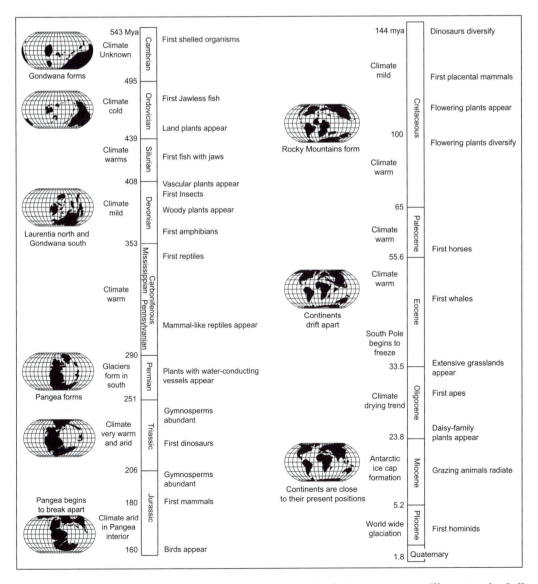

Figure 1.8: Geological time, illustrating forms of life and the shifting continents. Illustration by Jeff Dixon.

these divisions and subdivisions is identified by its characteristic fossils. Such *index* fossils define a geological time unit because they are widespread and common in rocks that appear in a certain stratigraphic position, but absent in rocks from higher or lower stratigraphic positions.

LIFE THROUGH TIME

Pre-Cambrian

The fossil record of life begins with fossils of bacteria from about 3.5 billion years ago. Chemical evidence of eukaryotes (cells like our own, with DNA protected in separate nuclei) appears 2.7 billion years ago. Single-cell fossils are all that has been discovered up to about 1.2 billion years ago, when what seem to be fossil tracks left by multi-cellular organisms appeared. Fossil amoeba (large one-celled eukaryotes) are known from 750 million years ago, and the first fossils of multi-cellular organisms appear about 600 million years ago, in the Ediacaran fauna of Australia and fossils from the Doushantuo Formation in China.

The Cambrian Explosion

At the beginning of the first period of the Paleozoic, named the Cambrian (Latin for Wales), we find a sharp increase in the variety and numbers of fossils. Trilobites appear early in the Cambrian alongside various worms. By the middle of the Cambrian a wonderful variety of animals can be found. A particularly fine sample of these, with rare evidence of soft body parts, is preserved in the Burgess shale of Yoho national park, in the Canadian Rockies.

This sudden appearance of so many kinds of fossils during the early Cambrian is often called the *Cambrian explosion*. Many hypotheses have been advanced to explain this abrupt appearance of so many new kinds of life in the fossil record:

1. The first is that the Cambrian explosion probably represents the point at which hard parts, like shells and teeth, became common. Earlier fossils of multi-cellular life have been found, such as the Ediacaran fauna. But they lacked hard parts, and consequently were rarely preserved: soft parts decay too quickly to be preserved except in very unusual circumstances. In fact, fossilization is rare for any organism. But it is far, far more common for organisms with hard parts. For example, paleontologists have been finding fossils of the hard upper shells of trilobites for centuries. But fossils of their soft underbodies weren't found until the twentieth century, and it took sophisticated preparation techniques and X-rays to study them. So in part, the explosion may be an illusion produced by the fact that earlier life was far less likely to be fossilized. In addition, the development of hard parts would have triggered an intense evolutionary arms race: animals lacking the protection of shells and the advantage of teeth would be quickly overwhelmed by their better-equipped competitors if they did not adapt quickly. Animals with hard parts could also make their living in entirely new ways—signs of extensive bioturbation, in which living things stir up the ocean bottom, first appear around this time as well. Hard parts could anchor

muscles and other tools allowing for more effective digging, as the animal hunts for food beneath the surface. So the first development of hard parts probably triggered a period of rapid evolutionary change.

2. The rapid development and spread of hard parts could only begin once multi-celled organisms had already arisen. A bitterly cold ice age preceded the Cambrian by about a hundred million years. During the ice age the entire earth may have been covered in ice. This harsh time (sometimes called snowball earth) was followed by a very warm, tropical climate. These dramatic changes may have helped set the stage for new evolutionary developments. There is also evidence that increased mineral erosion (which had been suppressed during the preceding ice age) raised calcium concentrations in sea water at the beginning of the Cambrian. This may have contributed to the development of hard parts by making the necessary raw materials easier to gather. Finally, oxygen levels in the atmosphere were increasing. This may have allowed larger animals to maintain adequate oxygen levels in their tissues.

3. There is also a *logical* point that's worth making here. Our classification system itself contributes to the appearance of an explosion. Almost all animal phyla—the most basic kinds of animals—arose during the Cambrian explosion. But the differences between the phyla at this point were smaller than they are now, and the boundaries were fuzzier. So the levels of diversity represented by these early forms of the basic phyla are not equivalent to the diversity we see today. The first steps towards today's diversity occurred at the very first branchings that separate the basic kinds of animal life today. These first steps occurred in or a little before the Cambrian. But this does not mean that Cambrian evolution must have raced along at speeds never seen since. When we assign fossils to higher taxonomic ranks, we trace back resemblances between fossil animals and modern groups, applying (as best we can) the categories biology provides today, to a world that was less clearly differentiated than ours. In the Cambrian, these basic lineages were closer than they are today, after 500 million years of further change.

This logical illusion arises at lower taxonomic ranks as well. The first mammals in the fossil record are from the late Triassic. But the variety of living things didn't suddenly and dramatically increase when these animals arose: at the time, the differences between mammals and their reptile cousins were far smaller than they are today. Paleontologists generally separate early mammals from their reptilian forbearers and cousins by one basic characteristic: the mammals' lower jaw is a single bone (called the *dentary*), while the reptiles' lower jaws include three bones. Animals on opposite sides of this divide in the Triassic were otherwise very similar to each other. The ones we call reptiles had a very large dentary bone, with the other two bones being much smaller. One group even had a double jaw-joint, noted in the name of one species, *diarthrognathus,* or two-jointed-jaw: one of the two joints links the dentary and the squamosal bone, just like modern mammals, while the other connects the quadrate bone in the skull to the articular bone in the jaw, like modern reptiles. This may sound awkward, but it worked fine—snakes today have double-jointed jaws as well, which help them swallow large prey whole. Given a particular definition of what separates mammals from reptiles, what we call mammals first appeared when some organisms first crossed that line. The reasons for drawing the line where we do have as much to do with taxonomic convenience as any particular merit in drawing it exactly there.

Another example of how the logic of taxonomy obscures relations between different species comes from our own family tree. The first animals that separated from the chimpanzee line and began the long series of changes that eventually led to humans were just another kind of early ape. The special traits that distinguish us from chimpanzees (and chimpanzees from us) hadn't developed when the split took place. To identify a kind of fossil ape as members of the human side of the split, we need to find that they have some specific (and probably obscure) skeletal trait that humans have and chimpanzees lack. Such traits include smaller teeth, heavier enamel on the teeth, and a skull shaped to balance on the top of the spine rather than hanging forward from the end of the spine.

The point is that taxonomy is *retrospective:* it uses hindsight to decide how to classify ancient organisms. We don't classify ancient life according to the taxonomy we would have used if we'd lived and done biology then. After all, this would be very hard to do. The fossil record isn't nearly as complete as we'd need it to be if we wanted to develop an entirely new taxonomy. And such a taxonomy would obscure the relations between groups of living things then and the groups that we find now. But we pay a price for the convenience and ease of using just one taxonomic system: groups that are very clearly distinct today get harder to tell apart as we look further back into the past. As we approach the point where they can barely be distinguished, we have to use increasingly arbitrary rules to tell them apart. We divide past life up among ranks that were developed to categorize present-day life. This ensures that our taxonomy is quite natural today: many differences separate the present members of different ranks. But when we apply the same categories to past life, we often have to make arbitrary choices in order to assign borderline cases to one or another present-day group.

The Paleozoic

Following the Cambrian, increasingly familiar forms of life emerge. Hints of life on land appear in the Ordovician and become a common feature of Silurian sediments. Fish also become increasingly common in the Silurian, and the first fish with jaws make their appearance later in the Silurian. The first sharks show up in the Devonian, along with early amphibians (the first *tetrapods*, or four-footed animals). The first large land plants also emerge during the Devonian: some paleontologists even refer to a Devonian explosion in plant life. In the Carboniferous, huge trees and rich plant life dominated the land. Deep beds of coal made of fossil plant material were laid down. In the Permian, a group of peculiar reptiles arose called *Therapsids*. Unlike other reptiles they had teeth with different shapes and a legs-under-the-body posture (instead of legs that sprawl to the side, as in today's crocodiles and lizards). Over time these reptiles developed an increasingly mammal-like skeleton. Later members of this group include animals we now classify as mammals.

At the end of the Permian, something drastic happened: a massive extinction, the most deadly in the earth's entire history (so far). The extinction marked the end of the Paleozoic and the beginning of the Mesozoic. Mass extinctions had occurred before, at the end of the Cambrian, and at the end of the Silurian, too. And more were yet to come—the most famous mass extinction, which killed off the archosaurs (dinosaurs, the flying pterodactyls, and the great sea reptiles) and many other groups (including the shelled, squid-like ammonites), came at the end of the Cretaceous. This extinction marked the end of the Mesozoic and the beginning of the Cenozoic. A further mass extinction seems to be underway now: how severe it will be is still unclear, and partly up to us.

There appear to be many different causes of mass extinctions. The Cretaceous mass extinction followed the impact of a huge meteorite or comet at Chixulub, off what is now the Yucatan peninsula of Mexico. The cause of the Permian mass extinction remains unsettled. Massive volcanic eruptions are one important candidate. Other mass extinctions may have been caused by drastic swings in climate and the effects of these swings on ocean currents. For example, a shift in currents could cover large areas of seafloor with anoxic water (water with very low oxygen content) from the deep ocean. This process appears to have contributed to extinctions at the end of the Ordovician and the Devonian, but may not have been present or important in the Permian extinction.

The Permian mass extinction marks the end of the trilobites, poster animals for the Paleozoic. It was also the end of the larger mammal-like reptiles that

The coelecanth is probably the most famous Lazarus taxon. These lobe-finned fish were well-known as fossils from the early to mid-Paleozoic, and believed to be long-extinct. But in December of 1938, a strange fish was caught off the coast of South Africa. Though he thought it would be inedible, the captain of the fishing vessel kept the fish because he thought the curator of the local museum in New London, South Africa might want to buy it. And he was right: Marjorie Courtenay-Latimer saw immediately that the fish was something special. Unable to refrigerate it, she took the fish to a taxidermist and had both the fish and its entrails preserved. She then sent a letter including a sketch and some scales that had come off the fish to Professor James L. B. Smith, a chemist with a serious interest in ichthyology. Dr. Smith, looking at the sketch and the enclosed scales, recognized the fish as a coelacanth. But to be certain, he arranged to see it in person. Unable to get away until mid-February, Smith finally arrived in New London on February 16. There, his high hopes were confirmed: there was no doubt that this was a coelacanth. Smith named the fish *Latimeria chamlumnae*, in honor of Miss Latimer.

Coelacanths first appear in the middle of the Devonian, about 380 million years ago. Coelacanths are a sub-order of the *Sarcopterygii*, which also includes lung-fish. *Coelacanth* actually means "hollow spine," referring to the hollow spines of the fish's vertebrae, through which nerves and blood vessels stretch. They swim by moving their fins alternately on opposite sides, in a motion reminiscent of walking. With their muscular, lobed fins they were close cousins to the fish that gave rise to the first four-legged land animals. Despite leaving no fossils (at least, none we have found) for over 65 million years, a group of coelacanths survived to the present. These survivors live in deep caves, emerging at night to hunt in shallower waters. One species has been found from South Africa to the Comoros islands, while another occurs in Indonesia. Both are considered endangered.

developed towards the end of the Permian. At the end of the Permian up to 95 percent of marine species went extinct, along with 70 percent of land based families. Crinoids, sea creatures related to starfish and sea urchins, were

Figure 1.9: A trilobite, a common type of arthropod of Paleozoic times. Image used with the kind permission of Arizona Skies Meteorites <http://www.arizonaskiesmeteorites.com/>. © Arizona Skies Meteorites 2007.

almost totally wiped out, but a remnant survived into the Triassic and later diversified again. Today sea lilies and sea fans, forms of crinoid, can still be found in deeper waters. They are the only surviving filter feeders amongst the starfish phylum, Echinodermata (spiny skinned). Two plant groups, the Cordaites (gymnosperms [naked seed] with long, strappy parallel-veined leaves) and the Glossopteris, became extinct at the end of the Permian. But the conifers and gingkoes (a different family of gymnosperms), which had developed during the Permian, continued on into the Triassic. During the Triassic they became the dominant forms of plant life on land.

It can be difficult to tell exactly when a particular species or family became extinct. We mark their ends, tentatively, by the youngest fossil found so far. However, there are cases of *Lazarus taxa,* which are genera or families that disappear from the fossil record for long periods of time only to reappear in very similar form millions of years later. This shows that organisms can be present for very long periods of time without leaving a noticeable trace in the fossil record.

The Permian extinction occurred at a time when all the continents were combined in a single land mass called Pangaea. This rare gathering of the continental land masses may have put a strain on shallow-water sea life, reducing the available habitat and triggering intense competition. A slow decline in the numbers and varieties of many kinds of life seems to have occurred during the later Permian, but there is also evidence of a sudden, punctuating event at the end of the Permian. A major flood basalt eruption at the end of the Permian formed the Siberian Traps, the largest such formation in the world. Occupying 2 million square kilometers, the traps formed when a mass of hot mantle material called a plume rose up through the crust. The eruptions continued for about one million years, at just about the right time to have caused the massive extinction that ended the Permian. Comparatively trivial eruptions that humans have witnessed in historical times have had drastic impacts, lowering global temperatures measurably. The effects of much larger eruptions going on continuously for up to a million years are almost unimaginable. Particles blasted high into the atmosphere would have cooled the climate. But this cooling effect could have been followed by massive greenhouse warming as the particulates settled while CO_2 and other gases that erupted from the volcanoes persisted in the atmosphere for long periods of time. Such wild swings of climate would have made life difficult for many species.

Less dramatic changes in climate may also have contributed to the extinction. A single massive continent, most of Pangaea was hot, dry, and very far from the sea. These extremes of climate made life hard. Many amphibians died out as the coal swamp environments they had thrived in dried up. Heat may also have stressed marine life. And some parts of Pangaea were over the pole, subject to extreme cold and glaciation. Glaciation would have stressed marine life not just by changing temperatures, but also by lowering sea level and reducing the shallow water environments surrounding Pangaea even further.

The Mesozoic

In the Triassic period, marine reptiles including icthyosaurs and placodonts appear. Early placodonts were lizard-like reptiles with flattened tails for swimming and strong, flat teeth for eating shellfish. Becoming more and more specialized, many later developed turtle-like shells and paddle-shaped limbs. The placodonts disappeared at the end of the Triassic, but the elegant, fish-like icthyosaurs continued throughout the Mesozoic, only to disappear at the end of the Cretaceous. A single branch of the ammonites survived the Permian extinction. They recovered a little during the Triassic, and diversified and spread immensely during the Jurassic and Cretaceous. Also in the Triassic, bivalve (two-shelled) mollusks like clams and oysters became the dominant bottom-dwellers in ocean sediments: life in the oceans began to look fairly familiar.

The first archosauromorphs (ruling reptiles) also appear in the Triassic. Sometimes called thecodonts (socket-tooth) for the bony sockets their teeth were set in, they gave rise to early dinosaurs, pterosaurs, and crocodilians, all with their teeth set into bony sockets in their jaws. The success of the archosaurs pushed the mammal-like reptiles and their relatives, the first mammals, into the background: the age of reptiles, the heyday of the dinosaurs, had begun. Late in the Triassic the first fossil of a fully fledged mammal appears—an animal about 10 centimeters long. By then, early dinosaurs as large as six meters had already evolved. Mammals spent the rest of the Mesozoic in the shadow of their larger competitors (though a Repenomamus fossil found recently in China suggests that larger mammals did occasionally turn the tables on small dinosaurs). The Triassic also ended with a mass extinction. This time up to 60 percent of species died out, and about 12 percent of families disappeared forever.

The Jurassic is the centerpiece of the Mesozoic, a time when familiar dinosaurs dominated the stage on land: giant sauropods, Apatosaurus (commonly known as Brontosaurus), Diplodocus and Brachiosaurus shared the landscape with massive carnivores like Allosaurus and Ceratosaurus (the first dinosaur ever named) and armored or plated and spiked plant eaters like Nodosaurus and Stegosaurus. The Jurassic Murchison formation in the western United States contains some of the richest dinosaur fossil beds ever discovered. Another famous Jurassic citizen was archaeopteryx. The earliest bird in the fossil record, archaeopteryx is very similar to the small carnivorous dinosaurs of the time. It is classified as a bird because of several features, especially its feathers and furcula (wishbone), but we now know that none of these traits is unique to birds: some small carnivorous dinosaurs had feathers and wishbones, as well as other traits that only birds have now. And unlike modern birds, Archaeopteryx had teeth, no bill, a long bony tail, and unfused vertebrae. At best an awkward and limited flyer, Archaeopteryx is a wonderful example of an intermediate between dinosaurs and birds.

Despite the appearance of these early birds, Jurassic skies were dominated by pterosaurs—flying archosaurs whose wings were made of skin stretched out along an enormously elongated fourth finger. Early pterosaurs had long tails, but late in the Jurassic short-tailed pterodactyls appeared; these forms survived the end of the Jurassic and continued flying up to the end of the Cretaceous.

In the Jurassic seas the fish-like icthyosaurs, pleisiosaurs with paddle-limbs and long necks and tails, and ocean-going crocodiles lived alongside modern-looking sharks and rays. Ammonites spread and diversified. Familiar-looking corals grew in warm, shallow waters, while reef-building sponges occupied deeper sites. Plant life on land was dominated by cycads, conifers, and gymnosperms like ginkgoes.

Figure 1.10: Archaeopteryx lithographica. The wonderful detail preserved in this fossil is due to the fine-grained lithographic limestone it was buried in. Used with the kind permission of Rock and Water LLC http://www.rockandwater.com/.

The Jurassic was followed by the Cretaceous. This period marks the first appearance of modern flowering plants (angiosperms), together with pollinating insects such as bees. Ants and butterflies also appear for the first time in the Cretaceous. Dinosaurs continued to dominate the land, with Tyrannosaurs and the smaller but still fearsome Deinonychosaurs as poster carnivores, along with the plant-eating ceratopsians (horned, rhinoceros-like herbivores including Triceratops and Styracosaurus), and duck-billed dinosaurs, called ornithiscians. Pterodactyls continued to dominate the skies—the giant Quetzelcoatulus had a wing-spread of 11 to 12 meters. It was probably only capable of gliding and soaring flight, but smaller pterodactyls flew actively, flapping their wing membranes. Birds became skilled flyers, though some, such as the fish-eating Hesperornis, still had teeth.

The end of the Cretaceous brought the most famous mass extinction of them all. The extinction is generally believed to have been caused by the impact of an asteroid or comet, though the detailed mechanisms involved are still being debated. The consequences of the impact certainly were catastrophic: world wide fires were triggered by debris that rained down all over the earth after the impact and left traces of ash in the KT (Cretaceous-Tertiary) boundary. The fires were followed by a long period of cold and dark, wiping out plant and animal life. The atmosphere may also have been poisoned by chemical changes triggered by the impact's massive fireball.

Scientists continue to debate whether the impact is really responsible for the extinction of the dinosaurs. Some argue that the variety of dinosaurs had declined over the last few million years of the Cretaceous.

Every land animal weighing over 25 kg died out in the aftermath of the impact. The great sea reptiles, icthyosaurs, pleisiosaurs, and mososaurs all died out. So did the ammonites. Many foraminifera (single celled plants and animals with elaborate skeletons) died out, and the pterosaurs met their end too. But many smaller animals, including amphibians, lizards, snakes, crocodiles, birds, and mammals survived. So did the new flowering plants, along with conifers and gingkoes and others. In the seas and lakes, snails, clams, corals and sea lilies, along with fish ranging from the ancient sharks to the most recent bony fish also survived. The survivors began to repopulate the planet, spreading out and adapting to the new opportunities opened up by the extinction.

The Cenozoic

The Cenozoic begins with the Paleocene. Mammals spread and diversified, becoming the dominant land animals. Mammals are divided into four basic groups: the familiar placental mammals (*Eutheria*), include cats and dogs, cattle and antelopes, bears, raccoons, bats, apes, monkeys, pigs, shrews, rodents, horses, elephants, rhinoceroses, and, of course, human beings. The marsupials include kangaroos and wallabies, opossums, koalas, wombats, and Tasmanian devils. The monotremes are the egg-laying mammals, including the platypus and the spiny anteater (*echidna*). Finally, the multituberculates, named for the bumps (tubercules) on their teeth, are the only major group of mammals that is now extinct. First appearing in the Jurassic, the multituberculates continued for over 100 million years, up to the Oligocene. They included burrowing animals, tree climbers, and gnawers, and seem to have lived much as many of the rodents who replaced them do today.

The earliest Paleocene mammals are long since extinct: Condylarths (largely omnivores and herbivores), and their descendants the Mesonychids (important carnivores of the Paleocene and Eocene, despite their hoofed feet). These groups left a wide range of successors, however, and during the Cenozoic, mammals have taken on a huge variety of forms, ranging all over the earth, the seas, and the skies.

Different kinds of mammals evolved on different continents. Today the monotremes are found only in Australia and New Guinea. The marsupials of the northern hemisphere became extinct during the Tertiary period (the first period of the Cenozoic). But in the southern hemisphere marsupials persisted, developing into many different forms. Some marsupials are obviously unlike any form of placental mammal: Australia's kangaroos, for instance. But some marsupials look a lot like familiar placental mammals—the thylacine or Tasmanian wolf looked startlingly wolfish. Nevertheless, thylacines arose altogether entirely independently of wolves: their teeth and skeletons, not to mention the pouches in which females carried their young, are unmistakably marsupial.

When North and South America were first joined by the Isthmus of Panama, many marsupials in South America became extinct. The opossum, on the other hand, became a marsupial success story, spreading widely into North America. In Australia, humans arrived as long as 60,000 years ago, bringing dogs with them. Placental rats had arrived before them, probably on rafts of vegetation. Since Europeans arrived in the eighteenth century, cats and foxes imported by European settlers have wreaked havoc on small native marsupials, including many forms of wallaby, bilbies, and the carnivorous quoles. Many are now endangered. European settlers on the island of Tasmania also exterminated the thylacine, the last large marsupial carnivore.

During the Eocene and early Oligocene, rodents replaced the old multituberculates. Familiar kinds of hoofed mammals made their first appearance, including early camels, tapirs, and rhinoceroses. The first whales, the first primates, and the first bats also emerged. Some familiar birds appeared during these times, including eagles, pelicans, and quail. Unfamiliar birds included giant predatory land birds like Phorusrachus, which stood up to 2 meters high with a massive, sharply curved beak. Though they lacked teeth and a long tail, these birds were reminiscent of the small carnivorous dinosaurs of the Mesozoic: fast two-legged runners with powerful legs and claws. The Eocene ended with another mass extinction. Though it was a comparatively small one, many important Eocene forms, including the massive titanotheres, died out.

Flowering plants continued to spread and increase in variety, coming to dominate the landscapes of the Oligocene. A common landscape that was missing from the earlier scenes we have been imagining is *grassland*. Grasses first appeared in the Eocene (recently some have been found in the Cretaceous, but they remained very rare until after the Paleocene). Early grasses lived mostly on the banks of lakes, streams, and other bodies of water, but in the Oligocene, they began to spread to higher and dryer environments. Horses, which first appear in the Eocene, began to spread and develop in different directions during the Oligocene. The even-toed hoofed animals (artiodactyls) became widespread, including the first true camels with ruminant digestion. This was an important adaptation for grazing and browsing mammals, allowing them to extract much more food value from plant cellulose. Early forms of dog (canids) made their debut as well, alongside a growing variety of primates and rodents, and the first elephants with trunks. Saber-toothed tigers, a star among fossil mammals, also arose in the Oligocene. Another Oligocene star was indricotherium, the largest land mammal ever: a giant hornless rhinoceros weighing up to 25 tons.

The Miocene epoch represents the peak of the age of mammals. Lasting roughly 18 million years, from about 24 to just over 5 million years ago, it is the longest epoch of the Cenozoic. Grasslands spread on land as the climate became drier, and kelp forests appeared for the first time at sea. Horses moved from forests to grassland, reached the size of ponies, and spread from North America to Eurasia. Deer and giraffes made their first appearance in Eurasia, and early elephants called mastodontids spread from Africa into Eurasia and North America. In South America, giant armadillos called Glyptodonts flourished. Early apes like Dryopithecus appeared.

In the Miocene seas we find strange creatures called Desmostylians, with projecting tusks and odd teeth with many cusps. Porpoises and whales swam alongside the giant shark *Carcharodon megalodon*, which reached lengths of 15 meters. Early sea otters appeared in the new kelp forests. The Miocene climate was generally warmer than both the Oligocene before it and the Pliocene that followed. But late in the Miocene, the Indian subcontinent, continuing its northward motion, made contact with Asia, beginning the raising of the Himalayas. This contributed to a general cooling of the climate at the end of the Miocene.

Late in the Miocene we find evidence of fossil apes with traits that may place them early in the lineage leading to modern humans. *Sahelanthropus chadensis,* dating back to between 6 and 7 million years ago, has canines that wear on the tips and not the sides, like later humans but unlike chimpanzees, along with a heavy jaw bone and thick brow ridges like those seen in much later hominid fossils. *Orrorin tugenensis,* a fossil ape dating back 6 million years, has thick tooth enamel, like humans and unlike chimpanzees. It is also claimed to have greater bone density on the bottom side of the shaft joining the thigh bone (femur) to the hip (pelvis). This claim is important because density patterns in bone reflect the forces acting on the bone—in

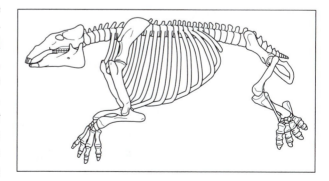

Figure 1.11: Desmostylian skeleton (Paleoparadoxia) prepared by Dr. Charles Repenning. After a drawing by Dr. Repenning for the USGS. Illustration by Jeff Dixon.

humans, the density is four times higher on the bottom side, but the density is the same all around the shaft in chimpanzees and other apes that walk on all fours. So the forces acting on Orrorin's femur may have been more like those acting on human femurs than on the femurs of modern apes. However, this claim is still under debate. Two later forms dating from 5.8 to 4.4 million years ago, *Ardipithecus kadabba* and *Ardipithecus ramidus,* are also candidates for early members of the hominin lineage.

The late Miocene cooling continued in the Pliocene. Ice caps formed over Antarctica and glaciers spread in the northern hemisphere. North and South America were joined together by the Isthmus of Panama, and animals spread both north and south across it—armadillos and ground sloths along with opossums and giant predatory ground birds moved into North America, while saber-toothed cats, wolves, and bears moved into South America, where they displaced the native marsupial carnivores. Land bridges also joined Alaska and Siberia, allowing mastodons to cross over into North America and three-toed horses to cross into Eurasia. Modern one-toed horses appeared for the first time.

Early members of the human family called Australopithecines (southern

apes) appeared in the early Pliocene. The climate in Africa was getting dryer, causing forests to retreat and open grasslands to spread. These changes may have made walking upright advantageous, but whatever the reason, it's clear that the Australopithecines walked on two legs. Early members of the group did retain some adaptations for tree-climbing, including long arms and slightly curved hands and feet. But their skulls were shaped to balance on top of the spine, and their hips and leg bones were also adapted for upright walking. The Australopithecines persisted for several million years ago; during that time, they became separated into two groups, the gracile (slim or light) and robust types (the latter sometimes grouped under the genus name *Paranthropus*). By today's standards they were all heavily muscled, but the robust australopithecines had much heavier jaws, teeth, and skulls. In height they ranged from as little

Figure 1.12: An Australopithecine skull. © Neanderthal Museum/ S. Pietrek

as a meter to as much as a meter and a half; males were considerably larger than females. Their brain sizes ranged from 350 cubic centimeters (comparable to today's great apes) to 550 (comfortably more than the great apes). Below the skull, later Australopithecine skeletons are somewhat more similar to modern humans in proportion and shape. *Australopithecus garhi*, one of the later gracile Australopithecines, appears to have made stone tools.

Homo habilis, the first species grouped in the same genus with modern humans, appeared about 2.4 million years ago. *Habilis* brain sizes range from 500 to 800 cc, with an average of about 650 cc. These figures approach *Homo erectus* at the upper end. Perhaps more important, some *Homo habilis* skulls may show signs of brain developments related to language, and used basic stone tools: sharp flakes of rock struck from a core with a hard hammerstone, these tools are collectively known as the Oldowan toolkit. Some experts divide *habilis* fossils into two groups, naming the larger-brained fossils *Homo rudolfensis*. Unfortunately, we know very little about the skeleton of these larger-brained individuals below the skull—evidence showing that their limb proportions and other features were becoming more human-like would do much to confirm their present assignment to the genus *Homo*.

Homo ergaster and *Homo erectus* remains have been found from as early 1.8 million years ago. Their bones were almost indistinguishable from modern humans,' except in skull shape: *Homo erectus* had a long, low skull with heavy brow ridges, and no chin. Their brain sizes range from 750 to about 1250 cc. At the

Dividing early humans into species is hard to do: the remains are usually incomplete, and we often have too few individuals to tell whether different-looking fossils belong to a single, varied, population, or to two separate species. Unfortunately there is no general solution to this problem. Such lines can be hard to draw even for species that exist today. For example, northern gulls are what biologists call a ring species. They vary quite smoothly, each type blending into and breeding with the next, in populations that are spread out around the north pole. But where the two ends of this ring meet, we find two separate populations that don't interbreed. Do these birds belong to a single species, or to two? If the intermediates died out, the extremes would clearly be separate species. But in fact, the intermediates link them, and if the extremes disappeared and an intermediate population survived, we would certainly regard what was left as one species.

We can't settle such questions by getting a ruling from some central authority: there is no such authority. Instead, these issues get sorted out over time, as individual scientists and groups of scientists argue for different views. Some scientists tend to be so-called splitters, interpreting fairly slight differences between fossils as marking different species. Others are so-called lumpers, more inclined to regard slight differences as variations within a single species. As evidence accumulates, opinions shift. For example, the Neanderthals were first regarded as a separate species from modern humans. More recently some anthropologists classified them as a (well-marked) subspecies, *Homo sapiens neanderthalensis*. But in the last few years, detailed studies of Neanderthal fossils (including DNA tests) have shown differences so great that most now classify them as a separate species.

Despite these debates, the fossils remain a solid touchstone for study and evaluation. They are the hard evidence on which our understanding of human evolution is based. Whether *Homo habilis* and *Homo rudolfensis* are two species, or just labels for individual differences between members of the same species, both the dates of these fossils and their intermediate status, in brain size and other features, between Australopithecines and *Homo erectus*, are clear.

Figure 1.13: Acheulean hand axes. © Neanderthal Museum.

upper end, this overlaps substantially with modern humans. *Homo ergaster* is slightly earlier and more primitive, but some experts regard these people as continuous with *Homo erectus,* and not a distinct species at all. Acheulean toolkits are associated with *erectus.* More complex and sophisticated than the Olduwan tools of *Homo habilis,* these basic stone tools included the *hand axe,* a large stone flaked to produce sharp edges and a point. The original, cruder form of the Acheulean toolkit is sometimes called the Abbeville kit. It dates to about 1.8 million years ago. The later, more refined version, first appears around 1.4 million years ago, and includes the hand axe, cleavers, picks, and other useful items. *Homo erectus* was the first human species to use fire, beginning at least 300,000 years ago. Marks on their bones where muscles attached indicate they were extremely strong. In a great migration, these people spread across Africa and into Europe and Asia. Imagine meeting these hunters on a lonely plain: a frightening thought!

Neanderthals—that is, *Homo neanderthalensis,* or *Homo sapiens neanderthalensis*—were named after the Neander valley in Germany, where the first fossils of these ancient humans were found in 1856. The Neanderthals inhabited Europe and the Middle East throughout the last ice age, up to about 30,000 years ago. Their average brain size was actually larger than that of modern humans. These were the first humans to bury their dead with tools and weapons at their sides, and ochre brightening their colors. Their tool-making techniques were rich and varied, though not as rich and varied as the modern humans who followed them. Neanderthals were heavily built and immensely strong, with long, low skulls, and beetling brows. At a height of about 170 cm, an average male is estimated to have weighed 100 kg (220 lbs). But despite their physical prowess, they were displaced by modern humans. It is still possible that some inter-breeding occurred, and even that some modern human genes derive from the Neanderthals—but so far there is no hard evidence to support such claims. The evidence we have suggests that the Neanderthals disappeared almost without a trace.

Almost 200,000 years ago, anatomically modern humans emerged in Africa. With a shorter skull, a higher forehead, and a strange projection on the lower jaw that we call a chin, these lightly built humans spread rapidly across the world. By around 55,000 years ago they had reached Australia. After sharing space with the Neanderthals in Europe and the Middle East, they became the only humans on earth some 30,000 years ago (with the isolated exception of *Homo floresiensis,* a still-controversial dwarf species from the island of Flores in Indonesia first reported in 2004; *Homo floresiensis* apparently persisted until as recently as 12,000 years ago). Crossing the Bering Strait sometime in the last 20,000 years, modern humans spread across North and South America, ranging from the lowlands to the high Andes, from the arctic to Tierra del Fuego. They colonized the islands of the Pacific, navigating the open ocean using outrigger canoes. Today, the entire world is their oyster.

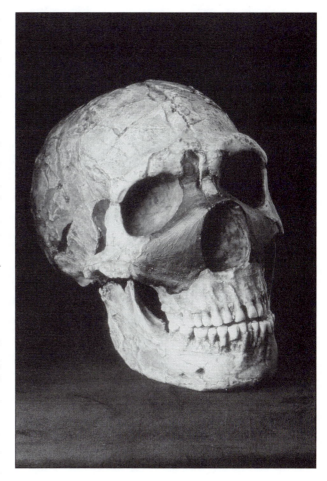

Figure 1.14: A Neanderthal skull. © Neanderthal Museum/ S. Pietrek

As they spread around the world, human hunters may have contributed to the extinction of many species: mastodons, mammoths, horses and camels, giant ground sloth, llama, giant beaver, glyptodont, giant peccary, woodland musk ox, saber-toothed cat, and cheetah all became extinct in North America very quickly, around the end of the last ice age. It may be that climate change played a role in these extinctions, but human hunters are widely believed to have contributed as well, perhaps dealing the final blow to populations already under stress.

Modern humans may also have contributed to earlier extinctions in Australia, Europe, Asia, and Africa. More recently, in New Zealand, the giant, flightless moas were quickly hunted to extinction after Polynesians colonized the islands. One striking fact about these extinctions is that they were particularly hard on big-game animals and their predators (such as saber-toothed cats). Both this and the timing of these extinctions seem to fit the hypothesis that human hunters had an important role in bringing them about.

Human populations have grown rapidly since the beginning of agriculture, about 10,000 years ago. Civilizations rose in the Middle East, in the Mediterranean, and in the Americas. Many fell later as their resources ran out, or as other civilizations conquered them. But today a complex and technologically powerful civilization reaches all around the globe to find the raw materials and energy that feed its economy. Human numbers have now surpassed 6 billion, and our impact on the world has grown to immense proportions. A mass extinction unlike any other in history is underway, the first to be brought about by the dominance and power of a single species. Like many other civilizations before it, our civilization is rapidly using up resources and damaging natural systems that it depends on. Whether we will be able to change course and find sustainable ways to continue remains an open question.

NOTES

1. The word *fossil* comes from the Latin meaning "dug up," and was originally applied to crystals and other objects found in the ground as well as to the traces of ancient life we now call fossils.
2. The full title was *De Solido Intra Solidium Naturaliter Contento Dissertationis Prodromus* (Forerunner of a dissertation of a solid naturally contained within a solid).

2

THE TREE OF LIFE

See, through this air, this ocean, and this earth,
All matter quick, and bursting into birth.
Above, how wide! how deep extend below!
Vast chain of being, which from God began,
Natures ethereal, human, angel, man,
Beast, bird, fish, insect! what no eye can see,
No glass can reach! from Infinite to thee,
From thee to Nothing!—On superior powers
Were we to press, inferior might on ours:
Or in the full creation leave a void,
Where, one step broken, the great scale's destroyed:
From Nature's chain whatever link you strike,
Tenth or ten thousandth, breaks the chain alike.

Pope (1734, Epistle I)

Nothing can be more hopeless than to attempt to explain th[e] similarity of pattern in members of the same class, by utility or by the doctrine of final causes. The hopelessness of the attempt has been expressly admitted by Owen in his most interesting work on the "Nature of Limbs." On the ordinary view of the independent creation of each being, we can only say that so it is;—that it has so pleased the Creator to construct each animal and plant.

Darwin (1859, 435)

LIFE EXAMINED

Beginning in the early nineteenth century, the fossil record of life's history allowed geologists to recognize and order different geological periods. Our first outline of earth's history was founded on the record of life's changes. But the history of life itself remained mysterious. That changes had occurred was clear, but many details were missing. More importantly, scientists did not know why or how the changes had come about.

Scientists like Georges Cuvier and Charles Lyell turned their attention to these questions. Was the tempo of life's history staccato (a process of sudden, large steps) or legato (continuous and smooth)? Did life change by sudden wholesale extinctions and replacements, by slow adaptation over time, or through occasional, piecemeal extinctions and replacements? Are living things today recently created species that replaced earlier, vanished forms? Or are they the descendents of ancient plants and animals, altered over the course of time?

There was also much still to learn about life around the world. Were the kinds of plants and animals found in each place a function of the climate and landscape as Buffon had claimed, saying "The Earth makes the plants; the Earth and the plants make the animals" (Buffon, 1749–1804, in Mayr 1982, 441)? Is the history of life the same everywhere, or did life develop differently in different regions? At a more fundamental level, scientists were re-examining basic questions about life. What properties of life do scientists need to explain? What kind of explanation can we give for them, and what can the history of life teach us about them?

FINAL CAUSATION

The scientific revolution of the seventeenth century underlies the last question. Newton's universe, the world of classical mechanics, had replaced Aristotle's metaphysics. Aristotle's world was full of natural purposes. For instance, Aristotle would explain an apple's fall towards the center of the earth by saying that, like all heavy matter, it was seeking a simple goal: to be *where it belonged*. But Newton's physics explained events mechanically, by appeal to what Aristotle called *efficient causes*. These causes act in the here and now, independent of any kind of purpose. For Newton the force of gravity acts on the apple, causing it to accelerate towards the earth's center of mass—the apple is not seeking any particular location, it's merely going where the forces of nature push it. Newton's theory was a scientific triumph: it also explained the orbits of the planets and comets, not to mention the tides. But in this new physical science, purposes and goals (the central elements of what Aristotle called *final causation*) had no place at all.

One important theme in this chapter is the idea of adaptation. A history of life is not just a history of things that have changed over time—it is a history of things with some very special properties. Living things in their ordinary environments survive and reproduce themselves, leaving very similar living things behind when they die. When we look at the details of their limbs and organs, we can see that they are *suited* to doing the things they must do to survive and reproduce. Plant-eating animals have teeth and digestive systems that they use to grind up and extract nutrients from the plants they eat. Carnivores have sharp teeth and claws that help them catch and consume their prey and digestive systems suited to extracting nutrients from meat instead of plants. Plants have special features that help them to retain moisture, attract pollinating insects and to protect and distribute their seed.

Because of these kinds of features, describing living things without appeal to purposes and goals seems ridiculous. Living things and their parts work together in complex ways. As Harvey showed in 1628, the heart pumps blood through the circulatory system. Lungs extract oxygen from the air, while the stomach and intestines extract nourishment from the food we eat. Plants' leaves use energy from the sun and gases from the air for nourishment, while their roots draw moisture and minerals from the soil. These organs work together to keep life going, while sexual organs enable living things to reproduce. Although nineteenth century scientists were just beginning to understand these organs and how they contribute to life, it was already clear that they carried out complex physical and chemical functions without which life would be impossible. The science of living things clearly needed an account of these organs and how they function. It also needed an explanation of how they came to be arranged so that each can serve its function, and how all of them work together to keep life going.

The obvious parallel was with human machines, which also have complex parts arranged to serve a purpose. In 1802, The Reverend William Paley published an influential book about purpose in life, titled *Natural Theology: or, Evidences of the Existence and Attributes of the Deity, Collected from the Appearances of Nature*. Paley argued that the intricate workings of living things made it obvious that they had been designed. His argument compared living things to a watch found on a lonely beach: having found such a watch, we would not hesitate to conclude that it had been designed and constructed for the purpose of keeping time. But Paley saw in living things the very signs of design that he saw in the watch—many complex parts, arranged in a very specific way that allows it to keep time when other arrangements would not—and, he argued, these features were present

W.Kidd del.   Prudhomme Sc

Figure 2.1: William Paley, naturalist and theologian. Courtesy Library of Congress.

to a much higher degree in living things than in any machine humans could make.

Paley concluded that living things clearly point to God as their creator—his aim was to show what we can learn about God the creator by a careful study of the creation. While he recognized that there appear to be cruelties and imperfections in life, he was still convinced that nature as a whole demonstrated God's knowledge, goodness, and power. For Paley these traits of God were most clearly revealed in the complex adaptations that enable plants and animals to survive and reproduce—their adjustment to the needs of these living things reflected a sort of perfection that pointed towards the perfection of their designer.

Georges Cuvier also saw adaptation as the key to understanding life. He noticed that different parts of organisms were linked by adaptation: each part makes a contribution to the organism's way of life. A carnivore, for instance, will have sharp teeth for killing and consuming its prey. But it will also have sharp senses and sharp claws that help it to detect and capture its prey. No animal combines the claws of a carnivore and the teeth of a horse or elephant. Cuvier believed that all the parts of an organism were shaped by its needs. The parts must all be integrated into a harmonious whole for the organism to succeed in life. Cuvier called this principle the *correlation of parts,* claiming that not only the obvious links between sharp teeth, claws, and acute senses, but also subtler links between the general structures of vertebrates and other basic groups of animals, were necessary for these animals to live where, when and how they do.

Cuvier argued that the correlation of parts limited how much a species could change. To go on being well-adapted to its way of life, a species had to stay pretty much the same. But Cuvier's work on comparative anatomy also showed that large groups of

Figure 2.2: Georges Cuvier, the great anatomist and paleontologist. Courtesy Library of Congress.

animals including species with very different ways of life still shared many complex structures. In fact it was these similarities that guided Cuvier's famously successful reconstructions of fossil animals from fragmentary remains.

As the greatest comparative anatomist of his day, Cuvier also embraced a second principle which was central to how he saw these relationships between different organisms. Called the *subordination of characters,* it held that the parts that are most important for correctly categorizing a species (i.e., for identifying its place in Cuvier's taxonomy) were the traits *least altered* by that species' particular way of life. This principle certainly helped Cuvier in his comparative studies, including both living and fossil forms. But to admit that there are constant traits connecting very different forms of life, traits that weren't altered by the special needs of each species, hinted to other biologists that the needs of organisms and Cuvier's correlation of parts were *not* enough to explain their similarities.

Whatever their explanation, Cuvier used these widespread similarities to group animals into four basic types that he called embranchements: vertebrates, mollusks, arthropods (which he called the articulates), and radiates (including jellyfish and other radially symmetrical animals). According to Cuvier, each of these groups shares the same basic structure. For instance, the vertebrates, including fish, amphibians, reptiles, mammals, and birds, are all built on the same basic skeletal plan. The arrangements of their organs were also very similar. Above all, Cuvier found that the nervous systems of his four embranchements followed a pattern that could be recognized and traced in every member of the group. Mollusks shared their own arrangement of the nervous system and other organs. The arthropods or articulates, including spiders, insects, scorpions, crabs, and crayfish, all shared a jointed exoskeleton resembling a suit of armor. Finally, the radiates shared a general shape, with similar parts radiating out from the organism's center.

These deep and widely shared resemblances raised a difficult question for people who thought adaptation was the key to understanding life. Why are so many complex details of organs and structure shared among all the members of each group? Different species in the groups often lived dramatically different lives, each in its own environment. They often had very different adaptive needs. So it was very hard to use adaptation to explain why they were so similar in structure. After all, there were large *differences* in structure between the different embranchements, which made it perfectly clear that alternative structures could also work. Further, animals from different embranchements often shared the same environment—for example, fish (vertebrates), jellyfish (radiates), and squid (mollusks) all swim in the same seas.

Still, Cuvier maintained that these similarities were adaptive. They were there to meet needs shared amongst the members of each group. Paley took the same position—both thought that each species was perfectly adapted to the particular kind of life it led, believing that further study would support this conclusion. Since it wasn't clear that these shared structures were needed by every member of the group, both Cuvier and Paley held that more work needed to be done to understand how these shared structures contributed to the life of the animals that possessed them.

RHYTHMS IN THE HISTORY OF LIFE

According to Cuvier, the history of life was a real thriller. He held that massive extinctions punctuated the history of life, when catastrophic geological events which he called *revolutions* exterminated all life across wide regions. Following such a revolution, similar but distinct forms of life would migrate in from distant areas unaffected by the disaster. This pattern of extinction and replacement fit well with the fossil evidence he and Alexandre Brogniart had gathered in the Paris Basin. However, there was a gap in this story. Migration could not explain how life on earth had changed over the long run, with older types dying out and entirely new groups arising. To complete his story, Cuvier needed an account of how new forms of life arose. But he had little to say on this question. That new forms did arise was clear, but how they arose was a mystery.

Cuvier seems to have thought that natural causes were responsible: astronomers did not invoke direct intervention by God to explain the motions of the planets, and biologists were similarly reluctant to make supernatural intervention part of their science. Rather than appeal to what they called primary causes, that is direct acts of God, they preferred secondary causes, natural processes operating according to laws of nature. But Cuvier had no account of what the natural processes that produce new species might be, though he clearly thought of them as carrying out God's intention of populating the earth with well-adapted forms: for Cuvier as for Paley, life was the result (whether direct or indirect) of God's creative purpose in the world.

Cuvier's famous student and protégé, Louis Agassiz, had a much bolder temperament. Agassiz proposed that there had been as many as 50 extreme catastrophes in which all life on earth was destroyed. He also proposed that after each catastrophe had wiped the slate clean, completely new collections of living things

Figure 2.3: Louis Agassiz, expert on fossil fish and the most famous early proponent of the ice age. Courtesy Library of Congress.

were directly created to re-populate the earth. Finally, for Agassiz the creation process involved a direct intervention by God in the world. But other scientists remained reluctant to accept a scientific theory based on primary rather than secondary causes.

The geologist Charles Lyell entered the debate as an opponent of Cuvier's revolutions. Lyell became a very important influence on Darwin's thinking, and, later, a friend and mentor. In his *Principles of Geology* (1831) Lyell argued for a new approach to geology. On his account, the steady, gradual processes we see at work today were the key to explaining the geological record. Above all, Lyell emphasized the damage done by failing to appreciate the vastness of geological time. According to Lyell, the catastrophists had ended up invoking vague, unknown, and extremely violent processes in geology simply because they had not allowed enough time for gradual, familiar processes to work. Worse, the catastrophists' appeals to violent processes that cannot be observed today damaged the predictive power and testability of their geology.

In one famous passage early in the *Principles,* Lyell used a historical analogy to argue that the greatest problems in geology can only be solved by allowing time for familiar geological processes to do their work:

> How fatal every error as to the quantity of time must prove to the introduction of rational views concerning the state of things in former ages may be conceived by supposing that the annals . . . of a great nation were perused under the impression that they occurred in a period of one hundred instead of two thousand years. Such a portion of history would immediately assume the air of a romance . . . armies and fleets would appear to be assembled only to be destroyed, and cities built only to fall into ruins.
>
> Lyell (1830–33, vol. 1 78–79)

Figure 2.4: Charles Lyell trained as a lawyer, but his passion for geology and gifts as a writer led him to become one of the most influential geologists of the nineteenth century. Courtesy Library of Congress.

The same, Lyell says, goes for geology. To compress the time is to alter all the normal operations of nature, and lose the link between the present and the past that makes the past understandable.

Lyell's work showed that allowing enough time for familiar processes to do the work produced straightforward and testable interpretations of geological features that other geologists had used vague revolutions and catastrophes to explain. Lyell's study of Mount Etna, the largest volcano in Europe, is a vivid example of his approach and its implications for geological time. Examining a valley cut deep into the mountain, Lyell showed that the great volcano had grown gradually, by means of a long series of lava flows no larger than those that had been observed during historical times. Assuming that these lava flows had occurred at the same rate as those recorded in history, he concluded that the mountain had been growing for millions of years. But the sedimentary rock underlying Etna contains many species that are still alive in the Mediterranean today. Though Etna was millions of years old, it was still very young by geological standards. The success of this interpretation of Etna together with the resulting expansion of geological time supported Lyell's doctrine of *uniformitarianism*, which held that present processes, operating at presently observed rates, could account for the earth's geology. There was, Lyell concluded, no need to appeal to revolutions or catastrophes in geology, and every reason to avoid them.

Lyell took his uniformitarian views to their logical extreme. He rejected any overall directional change in his geology: for Lyell, the earth had continued more or less as it is now through all geological time. An important consequence of this was that he, almost alone amongst the geologists of his time, argued against the fossil evidence for the history of life. He dismissed the apparent sequence of fossils as the product of an inadequate sample. For example, reptile fossils dominated the secondary era (the Mesozoic), while fossil mammals were common in tertiary times. But Lyell argued that this was just because we hadn't yet found the fossil mammals of the secondary. Lyell also suggested that long, slow shifts of climate might someday return the earth to conditions suited for earlier life forms. If and when they do, he suggested, ancient life forms might return as well: "The huge iguanodon might reappear in the woods, and the ichthyosaur in the sea, while the pterodactyle might flit again through umbrageous groves of tree-ferns" (Lyell, 1830–1833, I: 123).

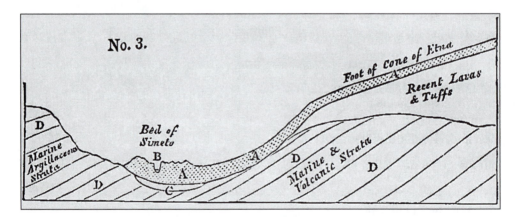

Figure 2.5: Mount Etna rising above recent Tertiary rocks. Image courtesy History of Science Collections, University of Oklahoma Libraries; © the Board of Regents of the University of Oklahoma.

Lyell and Cuvier agreed that new species arose over time by some unknown natural process. Both rejected the notion that God intervenes directly in the world, holding instead that God works through secondary, natural causes. Lyell hoped that a scientific account of these processes would eventually emerge. But like Cuvier and Agassiz, Lyell offered no account of the process that produced new species over time. The main disagreement between Cuvier and Lyell was over the tempo of the process: for Lyell it was gradual, with new species arising here and there, now and then. But for Cuvier, with his revolutions, the process could be sudden and involve the extinction (and the emergence) of many species at once.

ERASMUS DARWIN

One early evolutionary thinker was Erasmus Darwin, Charles Darwin's grandfather. He speculated about the possibility of evolution in his book, *Zoonomia: or the Laws of Organic Life,* which appeared late in the eighteenth century. His hypothesis was that evolution could explain the special adaptations of living things to their particular way of life. The forces behind this adaptive change, he suggested, were the drives of living things for mates, food, and security. Somehow, he proposed, these drives could bring about change in living things over time by their "own inherent activity." (E. Darwin, I, 572) As an example, he pointed to the different shapes of birds' bills, all adapted to suit the diet of each particular kind of birds. So Erasmus Darwin's evolutionary theory held that the demands of adapting to different ways of life can explain the special adaptations of different species.

Early evolutionary thinkers like Erasmus Darwin held that that life *improves* over time. The simplest version of this approach to evolution presents all life in a specific order on the steps of an evolutionary ladder from monad to man. The simplest and most primitive forms (sometimes called monads) are placed at the bottom, and humans at the top. This evolutionary ladder reflected the much older idea of the chain of being, a ranking of everything that exists, from the simplest bits of matter to God, on a universal scale of *perfection*.

The chain of being was often used to illustrate the *plenitude*, the fullness or completeness of creation: every link represented a tiny increase in complexity and perfection above the link below, and every position in the chain was supposed to be occupied by something.

Extinction and evolution both posed challenges to this picture. If life had changed over time, the idea of a chain that was complete and unchanging needed to be replaced. Evolution could preserve the chain, though, by treating the chain as a path that living things follow as they evolve towards perfection. But the chain of being does seem to rule out extinction: if some species had died out, then the chain could not be complete ("From nature's chain, whatever link you strike, tenth or ten thousandth, breaks the chain alike" [Pope, 1733]). The chain also posed a challenge to biologists: what is the right order, the order that links all life together in a single line, with one-celled organisms at the bottom and humans at the top? Neither Linnaeus's system of taxonomy nor its refinements by Cuvier and other comparative anatomists provided a single line along which all life could be ordered.

As we saw in chapter 1, Linnaeus organized life in a hierarchy: individual organisms belong to species, species belong to genera, genera to families, families to orders, and orders to classes. There is no way to arrange the species belonging to all these groups into one line running from the simplest to the most complex and sophisticated. Similarly, each of Cuvier's embranchements includes simpler and more complex organisms, but comparisons across very different groups are difficult to make.

LAMARCK

The most famous early evolutionist was named Jean Baptiste Pierre Antoine de Monet, Chevalier de Lamarck, though he was generally known simply as Lamarck. He began his career with a successful book on botany; after the French revolution he was appointed curator of insects, worms, and microscopic animals at the newly reorganized museum of France. Undaunted by having to master a completely new field of study, Lamarck set to work systematizing the huge variety of worms, insects, spiders, crustaceans, and mollusks. He was the first scientist to recognize that insects were fundamentally different from spiders and crustaceans. He also coined the term *invertebrates* to describe the whole range of animals he had been assigned to study.

Lamarck invented the term *biology* to describe the study of living things in general. Up to the late eighteenth century the general science of nature was called *natural history*. Natural historians studied everything from rocks and minerals to geology, meteorology, and plants and animals. But Lamarck saw that the study of life, which was distinguished from other subjects in natural history by the shared traits of growth, nourishment and reproduction, deserved a special name of its own.

Lamarck's account of evolution combined progress up the chain of being with the influence of adaptation. The result was a subtler version of the slow-climb-up-the-ladder picture of evolution. Lamarck held that life does have a general tendency to progress. But he also held that it tended to adapt to its particular environment and the opportunities that environment offered. This meant that evolution was not just a climb up the chain of being. Instead, life branches out in different directions as it evolves. Still, like other early forms of evolution, Lamarck's account made progress over time the central theme. Adaptation produced a complex, branching path, but evolution still moved continually upwards towards greater complexity and perfection.

Lamarck claimed that the simplest forms of life arise spontaneously. As they try to survive, their efforts alter their structure, improving them and adapting them to their environment: for example, ancestral giraffes may have stretched their necks reaching for food. When two such altered animals mate, the changes are passed along to their offspring. Although the general tendency of life is always towards greater complexity and sophistication, new life is always forming and beginning to evolve, so simple forms of life continue to exist.

Lamarck's theory of evolution allowed him to take a strong stand on another controversy. He rejected extinction, claiming instead that ancient forms of life have evolved into present forms life. This position made him very vulnerable. In a major public debate between Lamarck and Cuvier over evolution and extinction, Cuvier's case for extinction won out: many detailed, specific features of extinct animals, carefully observed and studied by Cuvier, have disappeared without a trace in today's animals. For instance, there are large bumps on mastodon teeth. They are completely different from any teeth known in living elephants. The structure of mastodon tusks was also very different from modern elephants. Lamarck had no credible candidates for the living descendants of the mastodon.

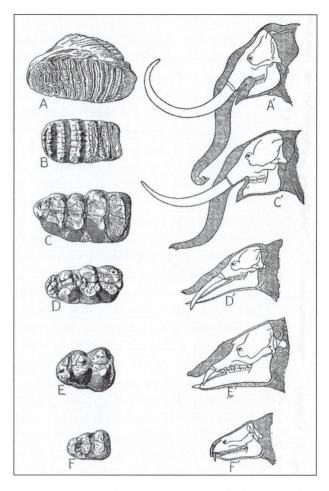

Figure 2.6: A mastodon tooth (C) and an elephant tooth (A). Note how different the two are. From Richard Swann Lull, *Organic Evolution*. The Macmillan Company, NY, 1924. Courtesy University of Calgary Library.

Evolution does link some past forms of life to their greatly altered descendants today. However, this is perfectly compatible with extinction: an evolutionary tree can have dead branches. Even if some species have descendants that are changed beyond recognition, others may have left no living descendants. Lamarck had added to his evolutionary views the further claim that every species in the fossil record has living descendants today. (Lamarck allowed a few exceptions, but these were all large mammals which he thought had been recently exterminated by humans.) But Cuvier's success in making the case for extinction also led most scientists to reject evolution, because Lamarck had linked his view of evolution so closely to his rejection of extinction.

GEOFFROY

Later, Cuvier won another famous debate against another evolutionist. This time, his opponent was Etienne Geoffroy St. Hillaire. Geoffroy was struck by some resemblances between vertebrates and mollusks. Generalizing on the resemblances, he came to believe in a unified plan covering all animals. According to this plan, all animals had the same basic components, with the same connections between them. In 1830, he presented a taxonomic scheme linking the invertebrates to other animals, following this principle of the unity of animal composition.

Geoffroy agreed with Goethe that nature includes a law of compensation or balancing of growth: when one organ develops further than normal, this occurs at the expense of other part(s). Geoffroy also maintained that nature takes no sudden leaps. Thus, even organs which are superfluous for a given species can be found in rudimentary form, if they play an important role in related species. These *vestigial* organs were evidence for the permanence of Geoffroy's unified plan: one example he used was the cassowary, a large flightless bird from New Guinea. The cassowary has vestigial wing bones, even though they have no use at all and lie completely under the skin of the birds. (In this respect they are a lot like the vestigial leg and pelvis bones found in some whales and snakes.) According to Geoffroy, there is no reason for the cassowary to have these peculiar internal limbs, aside from its close relationship with other birds. Conversely, the presence of these bones confirms the close connection between cassowaries and other birds.

Because the environment has varied, Geoffroy claimed that the forms of living things have changed since the origin of life. Still, he did not claim that existing species are changing today. So Geoffroy's evolutionary ideas had limited scope. More importantly, the idea that species might change in response to changing environments remained incomplete so long as no mechanism was proposed that could produce such directional change.

Cuvier argued vigorously against Geoffroy's unified plan. He held that his own four embranchements, representing the basic kinds of animals, were entirely distinct from each other. Drawing on the detailed anatomical studies he was famous for, he demonstrated that there were fundamental differences in both structure and development separating the embranchements. Using his principles, the correlation of parts and the subordination of characters, Cuvier could reconstruct fossil animals from just a few bones, and predict features of living species that hadn't yet been observed. So it's clear that Cuvier understood how closely the structures of animals within his embranchements matched each other. But he continued to insist that species were fixed. Each, he declared, had been created for the environment it naturally occupied, with each individual organ contrived with a view to its role in how that animal lives. Ironically, though Geoffroy lost the debate with Cuvier, recent work on the biochemistry of development has shown that some of the parallels Geoffroy noted between invertebrates and vertebrates really do point to a shared ancestor in the distant past.

Despite their differences, Lamarck, Geoffroy, and Cuvier all put the relation between environment and the organisms that live in it at the center of their biology—the *needs* of organisms, in a given environment, were thought to determine the features and patterns of development suited to those needs. The difference was that Geoffroy and Lamarck were gradualists: as the environment slowly changed, living things changed in response. As a catastrophist, Cuvier believed that geological changes were sudden and violent. So there was no question of gradual development of new species. For Cuvier, species were specifically created, according to the basic patterns of the embranchements, to fit with their environments. They could not adjust or change significantly without destroying the harmonious arrangement of parts that enabled them to succeed.

RICHARD OWEN

Richard Owen was the leading English comparative anatomist and paleontologist of the mid-nineteenth century. He coined the term *homology* in 1843, defining it as "the same organ in different animals under every variety of form and function." (Owen 1843, 379) He also invented the word *dinosaur,* or *terrible reptile,* for Megalosaurus, Iguanodon, and Hyleaosaurus, gigantic new fossils with reptilian traits. Like Cuvier, Owen was impressed by the similarities linking different groups. But unlike Cuvier, he did not claim that the similarities could all be explained by the adaptive needs of the animals. Instead, he endorsed a further law: the *unity of type.* Owen even produced a detailed account and drawing of the ur-vertebrate, which he imagined as an ideal animal possessing all the features shared by vertebrates, but none of the special features of particular vertebrates.

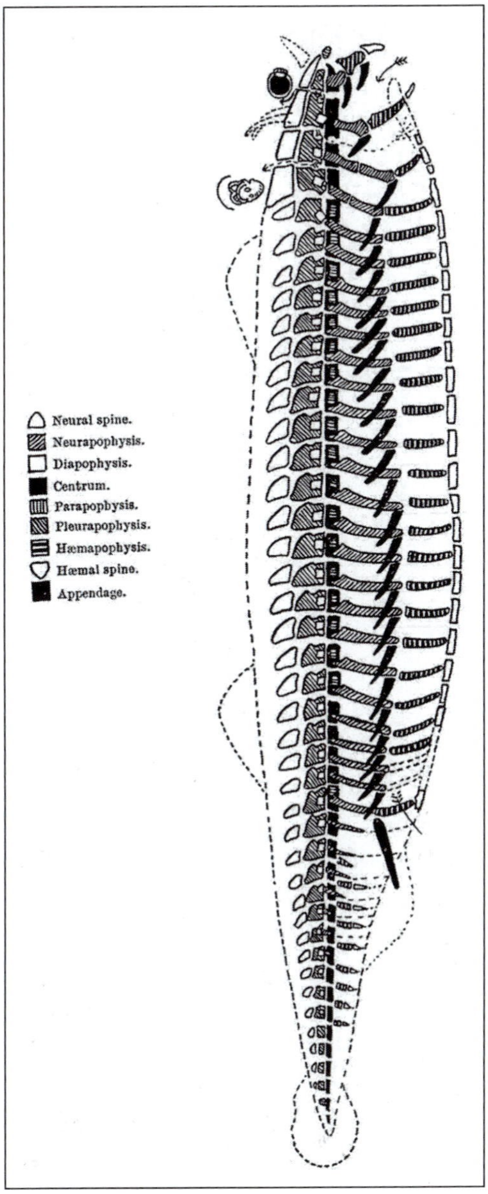

Owen did not claim that the vertebrates had all descended from a creature like the one he had described, though he did consider the idea. At first he proposed that a natural force like the force that shapes crystals set the basic patterns of development. Later, like the German biologists who had first conceived of such archetypes uniting different groups of

△ Neural spine.
▨ Neurapophysis.
▢ Diapophysis.
■ Centrum.
▦ Parapophysis.
▧ Pleurapophysis.
▤ Hæmapophysis.
▽ Hæmal spine.
▮ Appendage.

Figure 2.7: Owen's vertebrate archetype: an idea in the mind of God, or a representation of the common ancestor of all vertebrates? From E. S. Russell, *Form and Function: A contribution to the history of animal morphology.* 1912. Courtesy University of Calgary Library.

animals, Owen came to regard his ur-vertebrate as an idea in the mind of God, reflecting how God saw the relations between different groups of species.

Owen had taken a radical step by giving up the claim that adaptation was the key to understanding all the features of organisms. The central idea of Paley's natural theology was that final causation and the needs of the organism were the keystone of biology. But Owen's new law of the unity of type needed some other kind of explanation. By itself, the idea that his ur-types corresponded to God's general idea of each basic group of animals was a dead end, suggesting no further avenues of investigation or testing.

Charles Darwin was later to reject this view of Owen's, saying "[i]t is so easy to hide our ignorance under such expressions as the 'plan of creation,' 'unity of design,' &c., and to think that we give an explanation when we only restate a fact." (Darwin, 1859, 482) Darwin's point is simple: a scientific explanation must be the beginning of our investigations, not their end. Appealing to a plan of creation or to Owen's unity of type, contributes nothing to biology. It merely acknowledges the facts: we group living things under certain names, according to certain kinds of general and, so far, unexplained resemblances between them.

As to the origin of species, like Cuvier and Lyell, Owen believed that some natural process was responsible for producing new species, and offered no suggestion as to what that process might be.

EMBRYOLOGY AND DEVELOPMENT

More evidence of the close relations between different species emerged from the study of embryology. Karl Ernst von Baer conducted detailed studies, tracing patterns of animal development. His results reinforced the similarities of plan uniting Cuvier's embranchements. Though his work did not support Geoffroy's ambitious claims of a single structural plan for all animals, von Baer did recognize a shared *order* in the development of all animals. Growing from the single cell of a fertilized egg to the adult, every animal starts out as an undifferentiated clump of cells. As it grows, it gradually develops different tissues and organs. This process of differentiation begins with the most basic structures of its embranchement, and ends with the emergence of the special traits that identify its species. The undifferentiated starting point is shared by all multi-cellular animals; after that, development produces traits from the most general to the most specific in a set sequence.

Von Baer formulated four principles that have since come to be called *Baer's laws* for embryology:

1. The general characters of higher taxonomic ranks to which an embryo belongs appear earlier in development than the special characters of the lower ranks it belongs to.
2. The structural relations of the taxonomic ranks the embryo belongs to are formed in order: first those belonging to the higher (most general) ranks appear, then to the next level, and so on, until at last the most specific (that is, the structures characteristic of the species) appear.
3. Instead of passing through the adult state of other forms, the embryo, on the contrary, separates itself from them as it develops.
4. Fundamentally the embryo of a higher animal form never resembles the adult of another animal form, but only its embryo.

The third and fourth of these laws embody von Baer's rejection of *recapitulationism*. Recapitulationists held that embryos of higher forms pass through stages in which they resemble the adults of lower forms, and then go on from there to develop further. Some early evolutionists held that this recapitulation followed the same evolutionary path that the higher form's ancestors had followed as they evolved from the lower form to the higher. Von Baer replaced this linear view with a branching pattern of development which matched the branching pattern of taxonomy.

Von Baer's work also included a new and straightforward criterion for determining which forms were higher and lower. Lower forms had fewer different kinds of tissues, while higher forms had more. Von Baer labeled this measure the *grade of development*. Lower grade organisms were more homogeneous (in the very simplest cases they might have just one kind of tissue). Higher grade organisms were more heterogenous, with a wider range of more specialized tissues. Unlike the chain of being, these comparisons were confined to the embranchement that a particular organism belonged to. Even within an embranchement, it could only compare organisms that followed the same general path of development. For example, birds were neither higher nor lower than mammals on von Baer's scale. Unlike the chain of being, this new measure of grade fit comfortably into our branching taxonomy of groups within groups. Refinements in the details and new ideas about the base of the tree continue to emerge, but our general picture of life as arranged in a kind of tree remains the same today.

PALEONTOLOGY

Details of the history of life continued to emerge throughout the nineteenth century. As more fossils were discovered, Lyell's position became indefensible. The complete absence of mammals among fossils from the Mesozoic could no longer be explained away by claiming that the fossils found so far were a bad sample. At the same time, the work of comparative anatomists like Cuvier and Owen began to reveal interesting patterns in the fossil record. The fossil record had been recognized as the key to building a general geological sequence across Europe and even around the world. It was also clear that the plants and animals in the fossil record become more similar to living forms as we examine more recent fossils. But a subtler pattern of change was also beginning to emerge: not only had life changed over time, not only had it become more and more like modern life, it had actually branched out. Like taxonomy and embryology, the history of life also showed a branching pattern. General forms of each group appeared first in the record, while specialized forms emerged later.

This evidence was summarized in a prize essay by Heinrich Bronn in 1858. He confirmed the pattern of general forms first, followed by gradual specialization later, by a thorough analysis of the fossil evidence. Bronn found two general trends, one toward greater complexity in each lineage, and the other towards adaptation for increasingly specialized ways of life. The resulting picture was a branching tree, just as in taxonomy and embryology. But this time the tree's trunk represented the earliest, simplest, and most generalized forms of life. Each major branch left the trunk when the simplest and most generalized forms of the new type arose. New branches then grew off these branches in turn as more complex and specialized sub-types emerged.

Still, Bronn was not an evolutionist—his concern was only with the fossil record, and he could not find direct evidence there for transitions between species, despite the striking pattern of successive appearances of similar forms over time. The tree structure that Bronn recognized in the history of life is not the product of evolutionary thinking—instead, it served as independent evidence that Darwin drew on in his great argument for evolution.

By the mid-nineteenth century the branching pattern of life was evident throughout biology. In studies of living plants and animals, the hierarchy of species, genera, families, orders, classes, and phyla, reveals the branching pattern directly. In studies of embryos and how they develop, von Baer and his successors showed that the development of an animal embryo begins in the same way before following a branching path to its final adult form, as it develops the traits of its phylum, class, order, family, genus, and species appear in order. In paleontology, the same branching pattern appears all over again: earlier forms are more generalized, with more specialized forms appearing only later in the record. Any general theory of life needed to explain why the same pattern of branching appears in all three of these areas of study.

THE ORIGIN AND NATURE OF SPECIES

The question of how new species arise became more and more important as the list of known fossil species grew. At the same time, it was getting harder and harder for naturalists to say just what a species is.. This difficulty made the problem of how new species arose even more challenging.

For Cuvier, a species was defined by shared ancestry, which could be traced back to the first members of the species. Lyell thought in similar terms. But for naturalists in the field this definition didn't help: we can't tell whether or not a group of organisms have descended from common ancestors just by looking at them. More practical criteria had to be found. Worse, as naturalists made more detailed observations the question of where to draw the line got harder and harder: naturalists often could not agree whether two separate groups of organisms were just varieties of the same species, or different species altogether. Studies of hybridization did not settle the issue either: sometimes clearly distinct species were still fully fertile in cross-breeding.

Sometimes we recognize species by features that happen to be unique to them, or perhaps just unique to them in the region where they are found: if a medium sized North American songbird has a dull orange breast with brown feathers on its wings and back, it's a North American robin (a member of the thrush family). Simple identification rules like these have always been used by taxonomists, and they are still important for identifying organisms today. But when we try to construct a systematic taxonomy, we need to find the features each species shares with larger groups of other organisms, in order to assign them to the right higher taxonomic ranks.

Cuvier had emphasized that the key features for these groupings are the ones that haven't been altered in special ways to suit a particular species' way of life. The features we use to do taxonomy are the ones that are mostly unaltered throughout the wider group, the shared traits that Owen called homologies. These detailed similarities appear on many scales in both anatomical and developmental observations. They link the members of each genus, family, order, and class, no matter where or how the members of these groups make their living. But the bottom level of taxonomy is always the species level. If we cannot clearly identify the different species, our taxonomy is on very shaky ground. Taxonomists and paleontologists needed a new understanding of what a species is to help them do their work. Evidence and specimens from around the world were gathered and carefully compared before a new picture of the nature of species emerged.

BIOGEOGRAPHY AND LIFE AROUND THE WORLD

As exploration and conquest began to give way to (or include) scientific study, more and more information about life around the world became available. Scientific expeditions explored continents and islands, sending home specimens of thousands of new species of plants and animals. Traveling collectors, gathering both fossils and samples of new forms of life, provided new evidence of how life differs from one place to another.

One important contributor to this work was Charles Darwin, who traveled around the world on the HMS *Beagle*, a Cherokee-class Brig of the Royal Navy, between 1831 and 1836. Darwin's voyage, which laid the groundwork for his theory of evolution by natural selection, is the subject of the next chapter. His results, together with work by other young explorers, revealed yet another side to the branching pattern of life on earth: closely related forms appear in nearby regions, along with fossils of other closely related forms. Alfred Russell Wallace expressed the pattern very generally in 1855: new species arise close both in time and space to pre-existing, similar species.

3

DARWIN AND THE *BEAGLE*

PRELUDE

On the morning of December 27, 1831, a young man named Charles Robert Darwin put to sea on board a British Navy survey ship named HMS *Beagle*. He would not return to England until October 2, 1836. The main purpose of the *Beagle*'s mission was to draw up new charts of the South American coast and make improved measurements of longitude for some important locations. Darwin had joined the voyage as an independent naturalist, hoping to take advantage of the trip to study the biology and geology of the *Beagle*'s ports of call.

Darwin was also to be a companion to the Captain of the *Beagle*, Robert Fitz Roy. The young captain knew all too well the stress he would soon be under: Fitz Roy had served as lieutenant on the *Beagle*'s maiden voyage. During that voyage, his captain had broken under the strain and committed suicide. Fitz Roy returned to England in command of the vessel. Navy tradition forbade the captain to share his table or socialize with anyone of lower rank. Without a civilian gentleman on board, Captain Fitz Roy would have led a solitary existence for the duration of the voyage. Bringing a companion who could share his meals and speak with him informally was a wise precaution. It was also a clever plan to add to the scientific results and prestige of the mission—a plan that succeeded far beyond either Fitz Roy's or Darwin's expectations.

A recent graduate of Cambridge University, Darwin was well-trained for the role of naturalist despite his youth and inexperience. He had been born in Shrewsbury, England, on February 12, 1809, the son of a successful physician named Robert Darwin. Young Charles grew up in the English countryside around Shrewsbury, the main town of the county of Shropshire in the middle west of England. Energetic and outdoors-oriented, Charles was a mediocre student in school, where classics were emphasized instead of science. He

THE 'BEAGLE LAID ASHORE, RIVER SANTA CRUZ.

Figure 3.1: HMS *Beagle*: the vessel on which Charles Darwin learned to be a scientist. From Francis Darwin (editor), *The Life and letters of Charles Darwin, Vol 1. London: John Murray, 1888.* Courtesy University of Calgary Library.

began studies in medicine at the University of Edinburgh at the age of 16, thinking to follow in his father's footsteps. But Darwin found medical school unappealing; in particular he found the surgery demonstrations, which were conducted without anesthetic, unbearable to watch. Darwin gave up his studies in Edinburgh after two sessions. His disappointed father sent him to Christ's College at Cambridge University in 1828, encouraging him to study to become a clergyman. Darwin completed his degree in 1831.

During his time in Cambridge, Darwin enjoyed hunting with other well-off students. A little more unusual was his fondness for beetle-collecting, inspired by his entomologist-cousin, William Darwin Fox. One story from those days demonstrates how far he was willing to go in pursuit of a good specimen: while holding a beetle in each hand, Darwin suddenly spotted a third, of a kind he had never seen before. Rather than let go of the beetles he had already caught, he popped one into his mouth and reached out to catch the new one. Much to his surprise the beetle in his mouth immediately released a foul-tasting liquid. It and the new specimen both got away.

While at Cambridge, Darwin got to know two important figures: the geologist Adam Sedgwick, and John Stevens Henslow, a botanist and Anglican priest. He was seen so regularly on long walks with Henslow that students who didn't know Darwin referred to him as "the man who walks with Henslow." After finishing his degree, Darwin made plans for a scientific visit to the Canary Islands. In preparation, he read von Humboldt's account of scientific travels in the central and northern parts of South America; he also took a three-week trip to the north of Wales with Sedgwick to learn field geology. Humboldt was one of the great scientific adventurers of the age, and reading his book inspired Darwin with an ambition to become a scientist himself and to contribute to our knowledge of the world. Sadly, his plans for the Canary Island trip collapsed when the friend who was to accompany him suddenly died.

In search of a naturalist and companion for the *Beagle's* voyage, Fitz Roy wrote to Captain Beaufort, the innovative and scientifically minded commander of the Royal Navy's hydrographic office. Beaufort wrote to his friend Henslow, who suggested Darwin. Darwin was delighted when Henslow told him of this new chance to see the world and contribute to science. But Darwin's father refused to give his approval. Frustrated with his son's slow progress towards a

Figure 3.2: Darwin as a young man. From Donald Cultross Peattie, Green Laurels: The Lives and Achievements of the Great Naturalists. New York: Simon & Schuster, 1936. Courtesy University of Calgary Library.

settled life, Robert Darwin thought the trip would be a waste of time. Worse, he knew it would be uncomfortable, and it might be dangerous as well. But the senior Darwin left a loophole open: he told Charles that, if he could find a "man of good sense" who thought the trip a good idea, he would change his mind. Darwin quickly put the case before his grandfather, the famous industrialist Josiah Wedgewood. Happily for Darwin, Wedgewood thought the trip an excellent opportunity and interceded for him. Darwin's father relented, perhaps a little reluctantly. Darwin quickly contacted Fitz Roy to see if the position remained open. He was in luck: though Fitz Roy had offered it to a friend in the meantime, the friend had just declined. Darwin eagerly began preparations for the trip.

On Henslow's recommendation, Darwin brought the first volume of Lyell's *Principles of Geology* along on the journey. He studied it closely through the long days at sea, many of them spent in his cabin, seasick. Darwin soon found evidence of his own for Lyell's gradualist views: at the first landing of the voyage, on the Island of Santiago in the Cape Verde Islands, Darwin noted a long horizontal layer of white rock and seashells more than 40 feet above the present water level—a geological observation that led him to reflect on the processes that had lifted the layer so high above the water without breaking it

Figure 3.3: John Henslow, Darwin's mentor and friend. From Francis Darwin (editor), *More Letters of Charles Darwin.* John Murray, 1903. Courtesy University of Calgary Library.

up or tilting it. In Darwin's eyes, this was clear support for Lyell's gradualism: how could a sudden, violent process have lifted the layer so high without damaging it at all?

This idea of gradual elevation as an important geological process remained on Darwin's mind as the *Beagle* continued its survey along the east coast of South America to Tierra del Fuego and the Falkland Islands, and then around the tip of South America, through the Beagle channel and up the west coast to Chile and Peru. During a shore expedition on the coast of Chile, Darwin experienced a major earthquake that elevated the shoreline by about three feet along 300 miles of the coast. Lyell's gradualistic approach to geology was reinforced by Darwin's observations and ideas. He explained the raising of the Andes, one of the world's greatest mountain ranges, based on his observations of the earthquake he experienced. A long series of earthquakes, each no more violent that the earthquake he had witnessed, could have raised the Andes without the need for the violent, unprecedented catastrophes that Cuvier, Agassiz, and others had hypothesized. Similarly, Darwin saw that a gradual subsidence in the Pacific could solve the puzzle of how so many coral islands came to be dotted around the Pacific when almost no mountains rose above the water: as the original mountains subsided, the coral around them would continue to grow. Finally the mountains would disappear beneath the seas, leaving behind their growing coral crowns. Once again, a gradual mechanism worked without any need for sudden catastrophes.

Throughout the voyage, Darwin kept busy on shore whenever the *Beagle*'s schedule allowed. He traveled inland in South America on both the east and west coast, collecting plants, animals, and fossils, and making geological observations. He saw armadillos, rheas, guanacos, and capybaras, found fossils of giant armadillos, sloths, rodents, and mastodons, and gathered samples of new insects and plants, snakes, fish, and other marine life.

The specimens were carefully packed (though sometimes not carefully enough) and shipped back to Henslow at Cambridge. These large collections of new specimens were Darwin's first step towards a substantial scientific reputation. The writing he did on geological topics, including the Andes and his account of coral atolls, added to that reputation.

But our concern here is the impact the voyage had on Darwin's ideas about life. As we saw in chapters 1 and 2, three important facts were central to biology at the time. One was the relatively new idea that life has a history. Another was the growing recognition of the branching character of life, revealed in taxonomy, in the patterns of development and in the fossil record. The third was the importance of adaptation, the subtle and intricate features that suited living things to their different environments and ways of life.

THE *BEAGLE'S* BIOLOGY

Together with the work of other naturalist-explorers, Darwin's work on the *Beagle* added another important fact: that life is distributed around the world, not just according to the physical environment of climate and soils, but also according to regions. The fossils Darwin found, including the giant armadillo now named the *Glyptodon,* along with giant rodents and other specimens clearly presented a South American character: they resembled animals living in South America today—armadillos and capybaras, in particular—much more than animals living elsewhere. Many of the living species he sampled were also strikingly similar to other South American forms, though they differed sharply from species living in similar ways and in similar climates elsewhere in the world. It seemed that plants and animals from the same region tended to bear a kind of family resemblance to each other. In the tree of life, they were closer to each other than to living things elsewhere, even when they lived in very different environments.

What Darwin found at other locations reinforced this intriguing pattern. The most famous of these, of course, were the Galapagos Islands. The Galapagos are a group of volcanic islands right at the equator, 800 to 1000 kilometers west of the South American coast. They are cooler than their location suggests because the surrounding waters are part of a cold current flowing north from the Antarctic. The climate is very dry and the landscape harsh and bare, except on the higher slopes of the volcanic cones where more rain falls.

The *Beagle* arrived in the Galapagos on September 15, 1835, after surveying much of the Peruvian coast, where Captain Fitz Roy had left behind a boat and crew with orders to complete the survey of Peru and return home.

Figure 3.4: A Glyptodon—a giant, extinct cousin of the armadillo. From Rev. H.N. Hutchinson, *Extinct Monsters and Creatures of Other Days.* London: Chapman & Hall, 1910. Courtesy University of Alberta Library.

Figure 3.5: Armadillo. Fossil and living animals from a given region often show a 'family resemblance.' From Thomas Berick, *A General History of Quadripeds, 5th ed.* Newcastle upon Tyne: Edward Walker, 1807. Courtesy University of Calgary Library.

Darwin visited several of the islands, walking across the harsh landscapes, examining the many volcanic craters, and observing the wildlife. Marine iguanas basked in the sun on black flows of hardened lava, diving into the sea for meals of seaweed. Inland, other iguanas and giant tortoises crawled across the landscape eating cactus and other plants. The insects were dull-colored and few. The trees at lower elevations looked completely bare, but when Darwin examined them closely he saw that they all had very small leaves, and most were actually in bloom. The usual exuberance of the tropics was very subdued in the islands' dry, cool climate.

The absence of some kinds of animals was striking. The only mammal Darwin thought to be native was a mouse belonging to an American group—and he thought the mouse might have been recently imported to the islands. The other mammals he found, including rats and various domestic animals, were either deliberate imports or stowaways from the ships which had been visiting the islands for about 150 years. There were no frogs or other amphibians at all. Darwin explained this by contrasting the tough eggs of the iguanas, which just might survive a long sea journey from the mainland, with the delicate, soft eggs of amphibians, which clearly could not. Darwin concluded that the native wildlife on the islands had not arisen there, but instead was made up of descendents of plants and animals that had managed to survive the long, hard voyage from the mainland.

The seabirds were familiar, but the land birds were clearly special. Though related to birds from the distant mainland of South America, the species of the Galapagos were unique to the islands. Darwin collected mockingbirds, a hawk, two kinds of owl, and 13 species of the famous Galapagos finches, from heavy-beaked nut-crackers to slender-beaked insect-catchers. Like the insects, the birds were generally dull-colored, and in overall appearance and even tone of voice, they reminded Darwin vividly of South America and the pampas of Argentina. Darwin did not suspect, when he caught them, that the finches of different islands were generally of different species. As a result, he had mixed his specimens together—happily, though, this oversight was later corrected with specimens that other members of the *Beagle*'s crew had gathered and labeled by island. Darwin also didn't suspect that so many of the birds were finches, because of the wide range of sizes and the different shapes of their beaks. The insects on the different islands were also of different species. Darwin did not then grasp the importance of the fact that the shapes of the tortoise's shells differed from island to island, either.

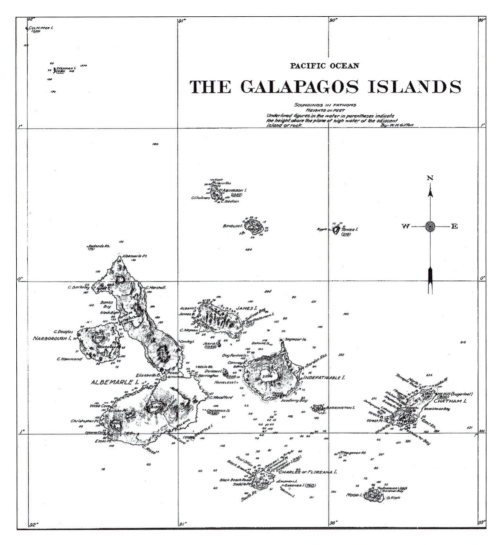

Figure 3.6: A map of the Galapagos islands. Each island is largely isolated from the others by winds and cold currents sweeping between the islands. From Alban Stewart, *Proceedings of the California Academy of Sciences, Fourth series, vol 1*. Courtesy of the University of Calgary Library.

Still, the idea that the islands had been populated by a very few species that had been lucky enough to survive the dangerous journey across the ocean was beginning to dawn on him. This implied that the new arrivals had subsequently changed and adapted to their circumstances, producing new, but closely related species. That mystery of mysteries, the origin of species, was near at hand in the Galapagos.

From the Galapagos, the *Beagle* continued its journey across the Pacific, visiting Tahiti, New Zealand, and Australia. The *Beagle* rounded the Horn of Africa on May 31, 1836 and put in to Cape Town. While there, Darwin met with Sir John Herschel, a famous astronomer and philosopher of science who had

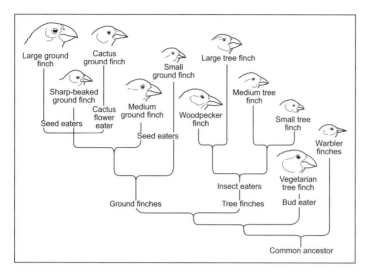

Figure 3.7: Relations between Galapagos finches and their beaks. Illustration by Jeff Dixon.

come to Cape Town to head the new Royal Observatory there. Darwin and Herschel discussed geology, biology, and ideas about the origin of species, including humans. From Cape Town, the *Beagle* went on to visit the barren island of St. Helena, briefly stopped again in South America to repeat some measurements, paused at the Azores, and then sailed for England. They arrived in Falmouth on the night of October 2. Darwin headed straight home.

After his return and a happy welcome home, Darwin set to work. He unloaded his last crates of specimens with help from Syms Covington, who had been his servant and assistant for most of the *Beagle*'s voyage and continued to work for Darwin afterwards. The crates were shipped to Henslow, whom he soon went to visit. There Darwin met Charles Lyell and Richard Owen. Lyell and Henslow advised Darwin on finding the right experts to study his specimens. Owen himself volunteered to study some of them. After a few months of travel and hard work, Darwin left Cambridge. He moved in briefly with his brother in London in early March, before finding rooms of his own in the city.

Figure 3.8: Marine iguanas in the Galapagos. These iguanas have adopted a new lifestyle, swimming underwater to eat seaweed growing offshore. From Archibald Geikie, *Textbook of Geology*, 4th edition. London: Macmillan, 1923. Courtesy University of Calgary Library.

TRANSMUTATIONISM

In late March of 1837, Darwin met with John Gould, an ornithologist who had been working on Darwin's bird specimens. Gould told him that all of the Galapagos finches were very closely related species, despite the variety of their beaks. Further, they all belonged to a genus unique to the Galapagos. Gould also pronounced the smaller variety of Rhea that Darwin had discovered in South America a separate species in its own right. Darwin's puzzling about species and their distribution took a decisive turn at this time: he came to believe firmly that species were unstable, that they changed over time, and in particular that these different species of finch were the descendents of a common ancestor that had managed to reach the Galapagos from the mainland long ago.

Darwin had long since accepted Lyell's geology, in which change was slow and steady but relentless. Each species lived in an environment that was gradually changing all the time. So a species could not stay well-adapted to its environment unless it could somehow change in response to the slow changes in its surroundings. Darwin also believed (though he later changed his mind about this) that Paley and Cuvier were right when they claimed that species were almost perfectly adapted to their environments and way of life. The only way to reconcile these ideas was to suppose that somehow, over time, species managed to adjust to the gradual changes in their surroundings.

In July 1837, Darwin began his first notebook on the transmutation of species. In it he drew a famous tree illustration, representing a single species that gives rise, over time, to several descendant species. He began to consider some important questions that were to guide his research for years: what evidence could he give in support of change in species? How did species manage to adapt to changes in their environment? How did new species arise from old ones? What could explain the rich similarities between species, similarities without which biological classification as we know it would be impossible?

If living things had been distributed around the world wherever conditions were right for them, it was strange that similar climates on different continents

Figure 3.9: Darwin's Rhea, a new species discovered by Darwin during his travels in South America. Linda Hall Library of Science, Engineering, & Technology.

should have such different plants and animals—and stranger still that the living and fossil species on the same continent should also show close resemblances to each other. The hypothesis of evolution offers an explanation of these observations. If living animals are descended from previous inhabitants, then it should be no surprise at all that some living species in a region resemble previous inhabitants that are now extinct. And if more than one species alive today were descended from a single ancestral species, then the existence of closely resembling species in a region is also explained.

Darwin was on the verge of becoming the first Darwinian. But a major question remained. Darwin was convinced that species changed, and that these changes enabled them to adapt, over time, to slow geological changes in their environment. But he had no account of what brought the changes about. Lamarck's theory of striving was teleological from the first. For Darwin this was unacceptable. A theory that explains adaptation by appeal to built-in teleology depends on teleology all the way down. No secondary cause of the kind that were so successful in physics, that is, no force that acts locally according to natural principles, can work that way. What Darwin wanted was a *mechanism,* a process that didn't depend on teleological assumptions. A satisfactory explanation demanded a non-teleological process that could produce adaptations, even though they seem so clearly directed towards a purpose.

In October of 1838, Darwin read a famous essay by Thomas Malthus. The essay advances a harsh thesis about human life, claiming that humans would always reproduce to the maximum numbers that could be supported. Our capacity to reproduce inevitably exceeds the available resources because reproduction produces *geometrical* growth, while at best the available resources increase linearly, and some may even be fixed. Suppose that a population of bacteria doubles every hour for as long as food is provided and waste products removed. In two hours we will have 4 times as many bacteria; in ten hours, 1024 times, and in 40 hours we'll have over a *trillion* times as many.

Any population that grows geometrically has a *doubling time.* If the population grows at 10 percent per year, then its doubling time is approximately 7 years (not 10, because as the population grows, the 10% annual increase grows too). Anything that increases so quickly soon outgrows the limited resources it needs to keep on growing. Darwin saw that Malthus's point applied to plants and animals just as much as to humans, and it focused his attention on the consequences of such growth when resources are limited: over time, some living things must fail to survive and reproduce. Darwin realized that the *struggle for existence* is an inescapable fact of life.

This melancholy thought was the central insight Darwin needed for his theory of evolution: in the struggle for existence, living things that have an advantage will tend to survive when others die. Any kind of feature could make a difference, depending on the challenges the population faced. Those who were a little stronger or a little tougher, who had a better immune system or faster feet, who could digest rougher food or had deeper roots to reach water in a drought, would sometimes survive when others could not.

Since many traits are heritable, some advantageous variations in a population would tend to spread through the population over time. Those with helpful traits would survive and reproduce more successfully. If their advantageous traits were heritable, their offspring would also tend to be more successful. So long as conditions consistently favored some heritable traits over others, the advantageous traits would spread, becoming more and more common. The population as a whole would evolve as advantageous traits that began as new and rare variations became common or even universal.

In his notebook, Darwin wrote "One may say there is a force like a hundred thousand wedges trying [to] force every kind of adapted structure into the gaps in the economy of nature, or rather forming gaps by thrusting out weaker ones" Darwin 1838, D notebook, in Barrett 1974, 456). "Here, then, I had at last got a theory by which to work," he wrote later (in F. Darwin 1887, Vol. 1, 83). In *The Origin of Species*, he outlined the argument forcefully:

> As many more individuals of each species are born than can possibly survive; and as, consequently, there is a frequently recurring struggle for existence, it follows that any being, if it vary however slightly in any manner profitable to itself, under the complex and sometimes varying conditions of life, will have a better chance of surviving, and thus be naturally selected. From the strong principle of inheritance, any selected variety will tend to propagate its new and modified form.
>
> Darwin (1859, 5)

Darwin was all too aware of the risk that being a transmutationist posed to his reputation. The idea that species change over time had been proposed many times before, but it had never won the allegiance of many working biologists. The idea that each species represented a single fixed essence was still dominant; variation was considered exceptional, and wild species were assumed to be almost perfectly alike. Worse, Cuvier had persuaded many biologists that there just wasn't room for species to change—as soon as a change became significant, it would interfere with the perfect correlation of parts that each species needed in order to make its living.

Darwin feared what might happen if his radical views about life were revealed. He saw clearly where his ideas led: the theory of natural selection explained the teleological aspect of life, so important to Paley's natural theology, as the product of a natural mechanism. The implications for God's place in a scientific view of the world were obvious. The biological world we find around us need not have been produced by a direct, purpose-imposing creation of life as we find it. Instead, given time and some simple form(s) of life as starting points, natural selection could produce the many sophisticated and varied forms we now find, adapted to so many environments and ways of life.

This view of life does not imply atheism. God's existence is compatible with evolution by natural selection. However, evolution does make our understanding of the natural world less dependent on theology, extending a trend towards purely mechanistic explanations of natural phenomena that had begun with the scientific revolution of the seventeenth century. Evolution is a natural process; it proceeds according to its own laws. Insofar as life is the product of evolution, no miracles are required to explain the biological facts. Neither do we need any revelations to study the history of life. We can test the evolutionary account of life in the same ways that we test any scientific claim about the past. The great power of evolution as an idea was clear to Darwin from the late 1830s; its success, in Darwin's time and since, has made it an indispensable part of how we understand the world.

Like many other thinkers of the time, Darwin regarded God's role in the world as limited, for reasons he was too modest to speculate about. Rather than intervene directly in particular events, God allowed the natural world to operate according to its own laws (the secondary causes of physics and other sciences). In 1860 he wrote to Asa Gray,

> I had no intention to write atheistically, but I own that I cannot see as plainly as others do, and as I should wish to do, evidence of design and beneficence on all sides of us. There seems to me too much misery in the world. I cannot persuade myself that a beneficent and omnipotent God would have designedly created the Ichneumonidae with the express intention of their feeding within the living bodies of caterpillars or that a cat should play with mice ... On the other hand, I cannot anyhow be contented to view this wonderful universe, and especially the nature of man, and to conclude that everything is the result of brute force. I am inclined to look at everything as resulting from designed laws, with the details, whether good or bad, left to the working out of what we may call chance.
>
> F. Darwin (1887, vol. 2, 311–12)

The discovery of a secondary cause that could explain the apparent goal-directedness of living things shifted God's role in the world another step further from the business of explaining the details of nature, towards a more remote and abstract role as creator. As many have observed, evolution and natural selection made atheism easier to defend. Rather than offer subtle philosophical objections to the argument from design as the philosopher David Hume had done, atheists after Darwin could give a direct reply: they could declare that evolution, rather than the hand of God, was the source of these designs. As Richard Dawkins has put it, Darwin's idea made it possible to be an *intellectually satisfied* atheist, one for whom the adaptations of living things are not a nagging, unanswered question but an expected and explicable part of nature.

PERSONAL EVENTS

Darwin's personal life changed at this time as well, in two ways. He had experienced several bouts of illness since 1837, going home to Shrewsbury and paying visits to the Wedgewoods to relax and recover. While visiting the Wedgewoods, he had formed a close bond with his cousin Emma. After thinking carefully about the advantages and disadvantages of married life, he paid another visit to the Wedgewoods in November of 1838. While there, he asked Emma to marry him. Against his father's advice, Darwin had already told her of his religious doubts, including his conviction that God does not intervene in the natural world. Happily for Darwin, Emma accepted his proposal anyway. Darwin moved into a house in London on January 1, 1839; he and Emma were married on January 29, and quickly settled in London. Their first child, a boy named William Erasmus Darwin, after Charles's grandfather, was born on December 27, 1839; a daughter named Anne Elizabeth followed in 1841.

By 1840 Darwin's illness had become a serious problem. He was often bedridden with migraines, chest pain, palpitations, and nausea. He soon found his busy life in London too much to cope with. The symptoms persisted, with some periods of quiescence, for the rest of his life. His father was unable to diagnose the illness; other doctors also tried and failed. Darwin's symptoms included anxiety, fainting, spots before his eyes, and ringing in his ears, eczema and blisters on his scalp. Some have suggested that the illness was brought on by stress over the radical turn in his thinking about evolution and worry about how his ideas would be received. Others have argued that he contracted Chagas' disease while traveling across the pampas of Argentina. He was certainly bitten by the beetle that carries the disease, a parasitic infection, and many of Darwin's symptoms fit well with a diagnosis of Chagas'. Other diagnoses range from psychosomatic illness or panic disorder to Ménières' disease and lupus. But the evidence is not conclusive and the case remains open.

Whenever he was well enough, Darwin continued to work. He sent a list of 21 questions about plant and animal breeding to breeders and farmers, asking about hybrids, crosses between wild and domestic animals and other issues. Unfortunately, few responded. On a visit to Shrewsbury in June and July of 1842 he prepared a short sketch of his theory, but continued to keep his radical ideas to himself. In July, after searching for about a year, Darwin finally found a suitable house in the country. The Darwins left London for Down house in the county of Kent that September, and Darwin took up the comfortable life of a country gentleman. Living quietly and keeping regular habits seemed to help, but he continued to lose months of work to recurrent bouts of illness. A second daughter was born in the fall; sadly, she died

Figure 3.10: Emma Wedgewood-Darwin. From Donald Cultross Peattie, Green Laurels: The Lives and Achievements of the Great Naturalists. New York: Simon & Schuster, 1936. Courtesy University of Calgary Library.

after just a few weeks. A third daughter, Henrietta Darwin, was born at Down house in September 1843.

In 1844 Darwin outlined his evolutionary ideas to Joseph Hooker, a botanist who had taken on the job of working through Darwin's plant specimens from Tierra del Fuego. Hooker soon became a close confidant, working as Darwin's research assistant in London. Darwin extended his sketch of 1842 to a longer essay of about 189 pages. Worried about his health, which had improved since the move to Down house but remained delicate, he prepared a letter to Emma asking her to publish and promote the essay if he should die unexpectedly. After editing by a local schoolteacher, Darwin hesitantly showed the manuscript to Emma that fall. Much to his relief, she raised few objections, although she did question some of his assumptions.

Figure 3.11: Joseph Hooker, one of the first biologists to learn of Darwin's ideas about evolution and natural selection. From Leonard Huxley, *Life and Letters of Sir Joseph Dalton Hooker*. London: John Murray, 1918. Courtesy of University of Calgary.

BARNACLES AND OTHER DISTRACTIONS

At this point, Darwin's theory employed natural selection chiefly to maintain the adaptation of species to their environments. As the environment slowly changed, natural selection would favor slow changes in plants and animals, adapting them to new conditions. New species would arise as migration and geological changes separated populations and selection slowly reshaped them to new environments and opportunities.

Robert Chamber's anonymous publication of *Vestiges of the Natural History of Creation* in 1844 marked an important shift in attitudes towards evolution. Aimed at a popular audience, the book was full of inaccuracies, which led most professional biologists to dismiss it. However, despite its poor reception in the scientific community, the book sold extremely well, effectively promoting its author's view of natural history as a process of progressive development according to natural laws among the wider public.

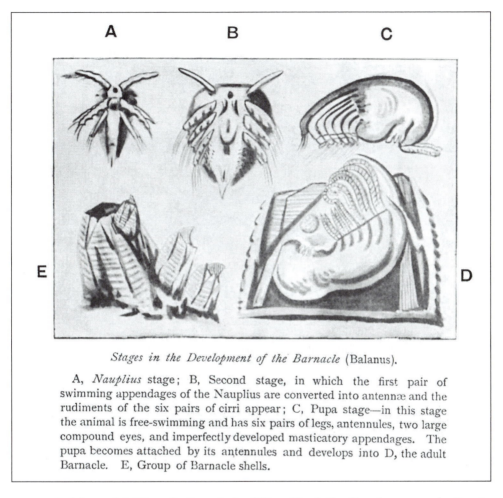

Stages in the Development of the Barnacle (Balanus).

A, *Nauplius* stage; B, Second stage, in which the first pair of swimming appendages of the Nauplius are converted into antennæ and the rudiments of the six pairs of cirri appear; C, Pupa stage—in this stage the animal is free-swimming and has six pairs of legs, antennules, two large compound eyes, and imperfectly developed masticatory appendages. The pupa becomes attached by its antennules and develops into D, the adult Barnacle. E, Group of Barnacle shells.

Figure 3.12: A barnacle's life cycle. From Arthur Milnes Marshall, editor, *Lectures on the Darwinian theory*. New York, Macmillan and Co, 1894. Courtesy of the University of Calgary Library.

Darwin's aim was not to write such a controversial, popular book. He wanted to persuade his colleagues in the scientific community that evolution by natural selection was the key to understanding life on earth. This was a much taller order. Despite the wide public attention it received, the hostile scientific response to Chambers's book confirmed that Darwin's work would face a tough audience.

Darwin's way of working through these worries was to continue reading and thinking about evolution from every point of view he could. Rather than publish immediately, he set about studying and taking notes on every field of biology, looking far and wide for evidence of common descent, for insight into the different kinds of adaptations living things possess and for information about heredity and variation. He took up breeding pigeons and corresponded with breeders, enthusiasts, and other scientists around the world. As he read new books on biology, he filled their back pages with short notes and page references for the ideas and facts that struck him as important.

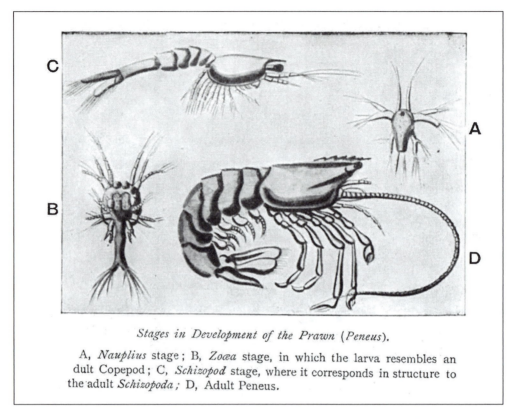

Stages in Development of the Prawn (Peneus).

A, *Nauplius* stage; B, *Zoœa* stage, in which the larva resembles an
dult Copepod; C, *Schizopod* stage, where it corresponds in structure to
the adult *Schizopoda;* D, Adult Peneus.

Figure 3.13: A prawn's life cycle. The first two stages of prawn and barnacle development are much
more similar than the later stages, where barnacles diverge dramatically from other crustaceans.
From Arthur Milnes Marshall, editor, Lectures on the Darwinian theory. New York, Macmillan and
Co, 1894. Courtesy of the University of Calgary Library.

All this time Darwin had also been busy writing up the results of the
Beagle voyage. The last bit of work left from the voyage proved fateful: Darwin began work on a barnacle specimen, which led him into an eight-year
long project. From 1846 to 1854 he studied barnacles in detail, gathering
specimens from collectors all over the world and writing four volumes on
their classification, their fossils, and their development, which takes them
from egg to free-swimming larvae to adults stuck on a rock or shell, or a
ship's hull. This work confirmed Darwin's status as an expert in biology, and
provided him with many detailed insights into barnacles that contributed to
his work on evolution.

Darwin and Emma's personal life included one major tragedy: the death of
their much-loved daughter Annie, at the age of 10. A bright, cheerful, active
child, Annie first fell ill in June of 1850; she took a turn for the worse in the
spring of 1851, and died—quite probably of tuberculosis—on April 23. After
Annie's death, Darwin found he could no longer believe in a personal, caring
God, or a special providence for each individual.

Darwin found telling anyone about his ideas difficult; he even wrote that
"it was like confessing a murder" (in F. Darwin 1887, vol. 2, 23). But over

the years, he took several colleagues into his confidence, revealing his ideas about evolution and natural selection to Joseph Hooker, Charles Lyell, Asa Gray, an American botanist, and to a few other friends. Their reactions were mixed. Some were won over, but most remained uncertain or even negative about Darwin's ideas. Still, Darwin's growing fame and reputation, cemented when he was awarded the Royal Medal of the Royal Society in 1853, ensured that he would get a hearing when he finally put them before the world.

Once the work on barnacles was done Darwin began what he called his "big species book," which he had been thinking about for so long. It was to be titled *Natural Selection*. The work went slowly—Darwin was carefully assembling the details of more than 15 years of steady work. But a major shock awaited him.

Figure 3.14: Asa Gray, an American biologist who corresponded with Darwin about his ideas before their publication. From Francis Darwin (editor), More Letters of Charles Darwin. John Murray, 1903. Courtesy University of Calgary Library.

4

ONE LONG ARGUMENT

As buds give rise by growth to fresh buds, and these, if vigorous, branch out and overtop on all sides many a feebler branch, so by generation I believe it has been with the great Tree of Life, which fills with its dead and broken branches the crust of the earth, and covers the surface with its ever branching and beautiful ramifications.

Darwin (1859, 177)

A SHARED DISCOVERY

A letter from the Molluca Islands reached Charles Darwin on June 18, 1858. Alfred Russell Wallace had written it that February, as he lay recovering from a bout of malaria. In the letter, under the title "On the tendency of varieties to depart indefinitely from the original type," Wallace proposed natural selection as the cause of the development of new species, the idea Darwin had been secretly working on since 1838. Darwin wrote immediately to Lyell, recommending that Wallace's paper be published, and lamenting that he had been forestalled by Wallace, though he resolved to continue work on his book all the same.

Lyell consulted quickly with Hooker, who had been among the first to learn of Darwin's ideas. Together they developed a plan for publication that gave both Darwin and Wallace credit for the new theory. Darwin would write up an abstract of his Essay of 1844, and both his paper and Wallace's would be presented to the Linnaean Society together. Despite hard personal circumstances—his infant son had just died—Darwin managed to provide a short abstract of his theory, drawn mostly from a letter he had sent to Asa Gray in 1857. The papers were duly presented to a small audience; publication in the Society's journal followed in August 1858.

Darwin abandoned the massive scholarly work he had planned, and started work on an abbreviated version of his Species book. The new book, titled

On the Origin of Species by Means of Natural Selection, or the Preservation of Favoured Races in the Struggle for Life, appeared in November 1859. Its impact more than made up for the underwhelming response to the presentation before the Linnaean society. Its first printing of 1,250 copies was sold out on the day it was released. A second edition of 3,000 copies with minor additions and corrections appeared in 1860. It soon sold out as well. A third edition of 2,000 copies appeared in 1861, with a new historical sketch of evolutionary thinking and a table of Darwin's revisions. The last edition revised by Darwin was the sixth, appearing in 1872.

Darwin described *The Origin* as "one long argument" (Darwin, Francis (ed.) 1887 Vol. 1, 103). It argues for evolution and natural selection, presenting evidence from every area of nineteenth century biology. The observations of naturalists, the practical experience of animal breeders, the observations and experiments of embryologists, the fossils discovered by paleontologists, the tree of life developed by the taxonomists, and studies of heredity and instinct all appear in the book. Darwin had built a thorough and carefully detailed case. The years he had spent in study and preparation ensured that his information was up-to-date and accurate; careful reflection and an open, honest approach to the views of his opponents are evident throughout. We'll outline Darwin's long argument briefly here, before going on to discuss how it was received.

THE ANALOGY BETWEEN ARTIFICIAL SELECTION AND NATURAL SELECTION

In the first five chapters of *The Origin,* Darwin introduced his theory using a parallel between the familiar process of plant and animal breeding, which he labeled *artificial* selection, and the novel evolutionary process he called natural selection. The first two chapters compare variation between and within domesticated breeds and wild species. Darwin had taken up pigeon breeding and corresponded actively with pigeon enthusiasts. Different breeds of fancy pigeons, he noted, breed true and are as distinct from each other as species or even genera in the wild. Yet they are all interfertile, and allowing them to interbreed produces, after just a few generations, birds that are indistinguishable from wild pigeons. He concluded that continued human selection based on small individual differences had produced animals as different from each other as distinct wild species, and it had done so in a very short time.

Heritable variation also exists in the wild—Darwin made a special study of variation in the barnacles he had lavished so much time on; he had notes and observations on variation in other groups as well. He knew that variations of as much as 20 percent in many features are common in wild populations. This provides a substantial base for selection to act on. Any heritable variation that provides an advantage in the business of survival and reproduction tends to persist and spread. Conversely, any heritable variation that reduces the chance of surviving and reproducing tends to disappear from the population over time, as its carriers die out and fail to reproduce.

Some advantages come and go—a drought, for instance, may select birds able to go longer without water, or able to eat larger seeds from the plants tough enough to keep growing when the soil is dry. But if the rains come back, thirstier birds that eat smaller seeds may make a comeback. Sometimes, though, an advantage has lasting power. A zebra colt that learns to run sooner after birth might escape predators that specialize in catching the young and vulnerable. And the same kinds of predators can be there, generation after generation, catching and eating more of the slower colts and fewer of the quick ones. If this difference between colts is heritable, it will spread through the zebra population until all colts are the quicker kind.

Of course which variations are advantageous for domesticated plants and animals depends on what human breeders are after. Perhaps it's disease resistance, or higher productivity, or cute floppy ears. So long as it's what the breeder wants, plants or animals that have it will be bred and those that lack it won't. Such artificial selection quickly produces changes that go well beyond the range of variation in the original population.

The parallel with natural selection is clear, but a little care is needed: in fact, Wallace objected to the phrase *natural selection* because it suggested a deliberate choice of some features over others. Wallace preferred Spencer's phrase, *survival of the fittest,* which Darwin included in some later editions of *The Origin.* Of course Darwin didn't think of natural selection as being carried out deliberately. The special status of survival and reproduction as goals makes deliberation and choice unnecessary. Nothing needs to seek these goals deliberately—*by the very definition of survival and reproduction,* any plant or animal that survives and reproduces successfully will leave descendents similar to themselves; those that don't (or that aren't as successful) leave no descendants (or fewer). This also leads to a substantial difference between the features natural selection can produce and the features artificial selection can produce. A feature favored by natural selection must have a history of contributing to the survival and reproduction of the living things that have it. Consequently, as Darwin saw, natural selection cannot produce organisms with features that only benefit some other kind of organism. For example, consider schmoos, the perfect domestic animal in Al Capp's comic strip, *L'il Abner.* Schmoos die of ecstasy if someone looks at them hungrily—and they taste wonderful, too. Perhaps such an animal could be produced by artificial selection. But they could not arise by natural selection, since this feature benefits humans at the expense of the schmoos' own survival and reproduction. Any feature like that would be selected against in the schmoo population, unless it was maintained by deliberate breeding. Careful handling would be needed, too: hungry farmers must stay away from the breeding schmoos!

Darwin also explained how the hypothesis of *descent with modification* from common ancestors accounts for the nested similarities that taxonomists use to group species, genera, and the higher taxonomic ranks. (Darwin takes this up again in greater detail, in chapter 13). The traits shared by a group are called *primitive,* meaning that they are inherited from the common ancestors of all the group's members, while the traits that distinguish the different species in the group are called *advanced* or *derived,* since they evolved in one line of descent from the common ancestors but not in the others. The history of these traits explains the nested structure of groups within groups that we find in taxonomy. Each branch in the tree of life is marked by new traits that arose in an ancestral population and were passed down to its descendants. By contrast, a design-based point of view provides no explanation of this pattern. Design can produce any useful combination of features. So design can't explain why so many groups lack features that would be useful for them—why, for instance, sea mammals like whales and seals lack gills. Neither can it explain why some features are so widely shared despite not being required for design reasons—for instance, the universal pattern of limbs in amphibians, reptiles, mammals, and birds: one bone closest to the body followed by two at the next step outward. Finally, design can't explain vestigial features, including teeth in the jaws of fetal baleen whales or the human appendix or the useless eyes of so many blind cave animals. But descent with modification makes perfect sense of them: they are not-yet-lost inheritances from ancestors for whom they were useful.

Darwin's Originality

Most of the ideas presented in *The Origin* are not original to Darwin. Both evolution itself (that is, descent with modification) and the idea that closely related species are descended from common ancestors had been widely discussed since the late eighteenth century. Natural selection had also been proposed in connection with some specific cases and as a way to keep organisms adapted over time. But it had not been put forward as the principle mechanism of evolution before Darwin and Wallace. The fact that life in general involves a struggle for existence had been widely noted. Heredity, of course, was also well-known, if not yet well understood. But the idea of sexual selection, which explains many apparently non-adaptive features of animals in terms of competition for mates, appears to be entirely Darwin's.

The most important aspect of Darwin's work that was truly original is the detailed work he did in building the case for evolution. Until Darwin, evolution had been an interesting and suggestive idea whose significance for biology had never been fully developed. The application of common descent to explain the distribution of species and groups of species, the groups-within-groups pattern of taxonomy and the branching patterns of development were all new with Darwin.

OBJECTIONS AND REPLIES

After introducing his theory using his analogy between natural and artificial selection, Darwin turned to address four difficulties that he knew would be raised. The first of these is the absence of transitional forms between closely related species. Even where the territories of two such species overlap, they generally don't blend together. Instead, the distinction between them is maintained. But Darwin's theory holds that the two groups descended from a common ancestor by a process of gradual change. Why, then, are they separated by a sharp line, instead of being linked by a continuous series of intermediate forms? Darwin's answer appeals to a number of factors: extinction, geological and ecological change, isolation, and competition. He concludes that intermediates linking the two did once exist, but they are now extinct. There are a few ways in which the disappearance of the intermediates could have occurred.

First, the two new species, having evolved by natural selection, must have had some advantage over the ancestral form that they replace. It may be that, having evolved separately because of a geological or ecological barrier, the distinct forms have since spread to the point that they overlap. On this scenario, the only intermediates were the original ancestors, now long extinct. Alternatively, having evolved in different directions in the two regions while still being linked by an intermediate population, the intermediates, a small, vulnerable population on the border of the present two territories, finally became extinct, so that the link between the present species was lost.

Darwin also offers a simple explanation, which he defends at length in chapter 9, of why we often have no fossils of these intermediates. Almost all dead organisms are completely lost—even elephant bones left in the open quickly break down and disappear. Fossilization is a rare event that generally depends on rapid burial in sediments. Further, in order for us to find the fossil today, those sediments must have continued to accumulate rather than being eroded and washed away, as sediments so often are. Consequently, the fossil record is extremely fragmentary and incomplete. We should only expect to find fossils of a few of the varieties of plants and animals that have existed.

The second difficulty Darwin discusses is the complexity and perfection of many adaptations. How could natural selection ever produce an organ like the vertebrate eye, with its lens, retina, optic nerve. and other parts all arranged to form an image useful to each organism? Darwin's answer is straightforward: it *is* hard to imagine, but there are many forms that eyes take in living things today, ranging from very simple light detectors to the complex eyes of vertebrates. These forms of eyes are distinguished from each other by quite small modifications. So long as each modification is useful in some way to the organism that has it, each could have been selected for as evolution continued. So, while we may have difficulty imagining the entire process, there is no reason to take that as evidence against evolution.

On the other hand, Darwin acknowledged that this imposes a limit on the kinds of features natural selection can explain. The adaptations natural selection can produce must follow this kind of gradual path. "If it could be demonstrated that any complex organism existed, which could not possibly have been formed by numerous, successive, slight modifications, my theory would absolutely break down" (Darwin 1859, 189). To show that such gradual paths exist, it's very helpful to consider some examples.

Darwin cites some surprising cases of gradual paths here, including the homology of swim bladders in fish with lungs in tetrapods, and of the gills of some barnacles with two small folds of skin in other barnacles, where they serve to hold the barnacles' eggs. He ends with a discussion of some difficult cases, including fish with electric organs and plants with sticky pollen packages. All these cases show a continuity of structure and development, though often the corresponding parts serve other purposes in related organisms. The concluding point is that over and over again, refined and complex organs essential to some living things turn out to have simpler parallels in the organs of other living things. Nature seems to make nothing entirely new, preferring instead to modify structures already in existence.

A related but subtler difficulty is the status of apparently unimportant organs. Dogs and cats get along perfectly well without their tails, so we may wonder why they have them at all. Would natural selection today favor the development of a tail in a kind of cat or dog that lacked one? If not, why do dogs and cats still have tails? Can natural selection preserve organs that aren't particularly important to the organisms that have them? On this issue, Darwin calls for a little modesty. It can be hard to determine just what use a particular feature has. Features that are initially selected for one function may later be altered to suit another, and unimportant traits can be linked to more important ones in ways that are not at all obvious. We are uncertain about why various domestic breeds differ from each other in the ways that they do, even though in these cases we know that the breeds are descended from a common ancestor and that they've been deliberately selected for certain traits. So it would be silly to require that natural selection provide an explanation for every detail of every organism. Even if they all are the result of evolution by natural selection, tracing that history in every detail will be difficult if not impossible.

Third, Darwin considers the origins of instinct. We call a behavior instinctive, he says, when it is done regularly, especially by young and inexperienced animals, and when it would require instruction or experience for us to learn to do something similar. The difficulty here is closely related to the worry about perfect adaptations already discussed: how could natural selection develop such an exquisite, complex pattern of successful behavior? As examples Darwin considers the slave-making instinct in some ants, and the honeybees' instinct for making the regular hexagonal cells of honeycomb. What is striking about the ants' instinct is that in some species the ants are completely dependent on their slaves, who do all the food gathering, all the tending of pupae, and who manage the moves of the colony, even carrying their masters in their jaws to the new nest site. Such species can become so dependent that they cannot even feed themselves without the assistance of their slaves. In the case of the bees, it seems that they have struck on a way of storing honey that uses the smallest possible amount of wax to store a given amount of honey. But it requires careful mathematical analysis to see that the honeycomb's structure manages to accomplish this. How, one wonders, could such situations arise? How could one species of insect become so completely dependent on the assistance of another? And how could bees have arrived at the best way of storing their honey using the minimum amount of wax?

Darwin's answers emerge from an examination of intermediate cases. Some species of ant do keep slaves, but are not so completely dependent on them. Also, their degree of dependence seems to vary from region to region—perhaps in relation to the availability of slaves. Darwin proposes a simple story, in which some pupae carried off as food during a raid manage to hatch, and the new-born worker ants go to work in the nest of their captors, performing the same tasks they would have performed in their own nest. The failure to eat all the captured pupae turns out to be beneficial to the nest, providing workers at a low cost. So selection favors this behavior, and a regular pattern of capturing slaves emerges. From this point, the availability and ease of capturing slaves together with the slave's ability to perform the necessary tasks in the nest will affect just how far a particular species may go in its dependence on slaves.

As for the honeybees, Darwin notes how bumble bees make roughly cylindrical wax cells for honey storage, attached to chambers where pupae had developed. He finds an intermediate in a Mexican bee which makes a rough sort of comb that has some of the features of honeybee combs, including flat shared walls where otherwise roughly spherical cells touch each other. He then considers evidence from detailed observations and experiments to determine more just how honeybees actually manage to make their comb. The result is an elegant account of how a more systematic and regular comb might arise, as individual bees work together following very simple rules for building comb. An important point here is that the total amount of wax required for storing honey will matter to natural selection: making wax demands a lot of energy from the bees; Darwin mentions an experimental figure of 10 to 12 pounds of dry sugar being required to form a single pound of wax. Since their food sources are limited, using as little wax as possible to store honey is very important to the bees. Any pattern of comb-building behavior that saved wax would be selected for.

The final item on Darwin's list of difficulties is the sterility of hybrids. This sterility takes two forms—first, it may be difficult or impossible to create a hybrid at all. That is, fertile plants and animals of different but closely related species are often not inter-fertile. The second form of hybrid sterility arises in the hybrid offspring, when parents of the separate species are inter-fertile. Mules are a familiar example—the offspring of a male horse and a female donkey, mules are tough, hardy animals (the result of the reverse cross is called a *hinny*), but they are sterile because chromosomal differences between horses and donkeys prevent mules from forming fertile sperm or eggs.

Darwin's worry here is that if sterility marks a sharp line between varieties and species, then we might take this to be evidence of a kind of impassable barrier between distinct species. But in fact the evidence is much more equivocal than this. The degree of sterility is difficult to measure reliably, but in general it varies widely. There are even asymmetrical cases in which males of one species are fertile with females of another, but females of the first are not fertile with males of the second. Further, naturalists sometimes base the line they draw between species and varieties on whether the two groups are inter-sterile. But this makes sterility between species a matter of definition, not a further fact confirming a special status for species. Darwin concludes that sterility arises as a result of other changes that have accumulated over time, rather than as a special property that is somehow selected for or imposed on species.

DISTRIBUTION OF PLANTS AND ANIMALS

In chapters 10 through 12 of *The Origin*, Darwin discusses the history and geography of life. Chapter 10 focuses on the pattern of extinction and appearance over time for species and higher ranks. Darwin's main point here is to establish that species and higher ranks tend to come and go in much the same way over time. From their initial appearance, they tend to spread and expand relatively rapidly, before gradually fading away. These appearances and disappearances have taken place throughout geological time, as Bronn's famous review essay on the fossil evidence showed; Darwin appeals to this evidence to support his gradualist view of how life on earth has changed over time. In particular, the evidence of appearances and disappearances doesn't support the pattern of mass extinctions followed by complete replacements that Louis Agassiz favored.

Another point Darwin makes here is that extinction should not surprise us at all. It's a commonplace observation that some species are rare. Whatever the reason for their rarity, such species are vulnerable. A new disease, a change of climate, or the arrival of a new predator on the scene can tip the balance against them once and for all. And when a species or genus or family is lost, it's lost forever. Similar forms might emerge later, but these new forms will always differ from the original because they have different ancestors from which they inherit different traits.

Chapters 11 and 12 examine the distribution of life around the world—Darwin's work on the *Beagle* voyage provided a lot of evidence here, but it is supplemented by the work of many other naturalists around the world. The role of geographical barriers in dividing life on earth into characteristic types and groups appearing in different regions is emphasized. For instance, Darwin points out that the Atlantic and Pacific sides of the Isthmus of Panama have entirely different shellfish, and the shellfish change completely again when you go further west to the Pacific Islands—but some particular forms range all the way from there to the east coast of Africa, almost halfway around the earth. The obvious explanation is that the distances from each island to the next island or coastline as we move from the Pacific islands to Africa is never great enough to be a barrier to migration.

Chapter 11 closes with a discussion showing how the recent ice age explains the uniformity of widely separated mountain species in the northern hemisphere. As the ice age began, arctic species would have migrated south to occupy areas abandoned by their former inhabitants, who had to migrate southwards to escape the cold. But arctic species are very uniform, because the land around the north pole is nearly continuous—with the lower sea levels of the ice age, there was a land bridge across the Bering Strait that separates Asia from Alaska, and the distances from North America to Greenland, Iceland and on to Europe in the far north are also very short. So the same plants and animals spread south from the arctic across Europe, Asia, and North America.

As the ice and cold retreated northwards at the end of the ice age, many of these species left remnant populations high in the mountains, where the cold conditions they depend on still persist. The result is that today, in mountainous regions across the northern hemisphere, the very same species can be found across great distances even though they could not migrate from one mountain to the next in the present climate.

Darwin also reports some experiments and observations on the migration of plants from one region to another. He studied how long the seeds of different plants could survive being soaked in saltwater. He also examined birds as a possible means of transportation for seeds. He looked at bird's feet to see how much soil and seeds they carried, he tested seeds from the crops of dead birds to see how long they could survive as the birds floated in saltwater, and he planted seeds from bird feces and owl pellets to see if they would germinate. He also examined tree roots and icebergs as means of transportation for seeds: a tree that is toppled over and washed into the ocean can carry large amounts of soil in its roots and icebergs regularly carry large amounts of soil and rock as well.

Some puzzles certainly remained: Darwin's friend Hooker had observed that similar species of plants occur very widely through the southern hemisphere. The question of how this had come about continued to be a difficult challenge for biogeography during the next 100 years. It was only in the 1960s that plate tectonics (the modern theory of continental drift) offered a simple explanation: the widely dispersed continents of the southern hemisphere were once actually connected to each other. Plate tectonics also helped with many other puzzles in the distribution of both present-day and fossil forms of life. For example, some plant and animal fossils from southern Africa and South America match each other perfectly; today we explain this by noting that, at that time, southern Africa and South America were actually joined together.

The general lesson of these chapters is that climate and physical environment are not the key to life's distribution. (This goes against what many of Darwin's predecessors had claimed, namely that each form of life was originally placed in the environments it was best suited to.) Specifics of location and geographical barriers to migration tell us much more about where species will and won't be found. If each species originates in a particular place and time, and (if successful) spreads out from there, then the distribution of life in space and time makes sense. As expressed by Wallace's law, that "every species has come into existence coincident both in space and time with a pre-existing closely-allied species" (Wallace 1855, 186), the support this gives to evolution is obvious: these patterns in the distribution of life through space and time are just what Darwin's theory of the origin of species through descent with modification would predict.

MORPHOLOGY, EMBRYOLOGY AND DEVELOPMENT, RUDIMENTARY ORGANS

Chapter 13 examines the branching pattern that appears in the similarities of different living things, in the patterns of their development, and in the persistence of certain structures or organs even in living things that have no apparent use for them. Once again, Darwin explains the branching pattern by descent with modification. On his account the shared features of each group have been inherited from common ancestors. This also accounts for why so many organisms lack traits that would be useful for them given their present ways of life. Those useful traits are restricted to groups descended from their first inventors, and time and the available variations have not (or not yet) enabled similar traits to develop in every organism that could benefit from them.

The main aim of the first part of the discussion is to show how the rules and patterns that scientists use to group species, genera, families, and all the higher ranks together are explained by evolution. The real link that ties all these groups together, Darwin argues, is descent from a common ancestor. This explains the shared characteristics by a simple appeal to inheritance, "the only certainly known cause of similarity in organic beings" (Darwin 1859, 456). It also explains many details. Why are *different* features important for recognizing members of different groups? Because the traits that are preserved through inheritance in one group need not be the same as those preserved in another group. Why are the features that reveal these relations often physiologically unimportant? Because these features are not as strongly subject to selection, and so are often preserved while more important features are changed. Why are vestigial organs so regularly observed and so often helpful in deciding the relations between different groups? Because they are the result of inheritance alone. Why are some features (the streamlined shapes of whales, fish, and dugongs) unimportant as clues to relations between these groups, but do indicate relations within them? Because fish, whales, and dugongs did not inherit their streamlined shape and fins from one and the same common ancestor, although the members of each group did inherit their streamlined shape and fins from a common ancestor. Why do all barnacles with complex mouths involving many parts have fewer legs than barnacles with simpler mouths? Because they are descended from barnacles with simpler mouths and more legs, and evolved their complex mouths by modifying some of their legs. Anatomical and developmental observations directly support this relation between mouth parts and legs in barnacles.

The branching pattern of taxonomy is repeated in development as well, as we noted earlier. One important question here is why evolution by natural selection would produce this branching pattern in development. Why wouldn't selection produce changes in the early stages of development as well as in the later ones? Why is natural selection so *conservative* with respect to development? After all, there is no reason why variations shouldn't arise at all points in the course of development. Is there a general reason why such variations would not be selected for?

Darwin offers an answer that begins with two general principles and then links these principles to evolution by natural selection. The principles are simple: first, most variations don't show fully until the variant organism reaches adulthood; Second, the variations that are inherited appear at the same point in development. When these principles are combined with the idea of descent with modification, that is, evolution, we would expect that adult organisms descended from a distant common ancestor will tend to become very different from the original adult form, while their young remain quite similar to the young of the original form. This in turn implies that when different species (or genera or families) have descended from a common ancestor, their young will tend to resemble each other more closely than the adult forms do. That is, embryology will be very useful, as indeed it is, to sorting out the relations of different forms.

Darwin points out that "(t)he points of structure, in which the embryos of widely different animals of the same class resemble each other, often have no direct relation to their conditions of existence" (Darwin 1859, 439–40). As an example, Darwin considers the vertebrates, all of which go through a stage in which arteries in the neck follow a loop near the branchial slits. This can't be due to some similar condition of existence that all vertebrate embryos encounter, because those conditions are so different: mammals develop in the mother's womb, birds in eggs in a nest, and frogs in eggs underwater. But they all develop branchial slits, and all have arteries that form a loop near those slits. So this is a trait that they all share and which we cannot explain as the result of natural selection for some advantage that it provides to the developing embryo. Since this feature is not the result of natural selection acting on each separately, it's best explained as the result of inheritance from a common ancestor that had the trait.

Natural selection also helps to explain both the general rule that variations don't appear fully until adulthood, as well as some exceptions to this rule. Darwin notes that having a long beak as a nestling is not important to a bird that uses a long beak to obtain food as an adult. Thus variations that tended to produce a long beak only as it approached adulthood would be at least as favorable as variations that produce the long beak earlier. On the other hand, if the young have to fend for themselves, and if their way of life is very similar to the adults' way of life, then we would expect the young to take on adult forms very quickly. Spiders and cephalopods (squid, octopuses, and their relatives) both show this pattern of development.

At this point, Darwin turns to evidence from dogs, horses, and pigeons. His aim is to extend the analogy between artificial and natural selection by applying it to embryology. Darwin had careful measurements made, which showed that for greyhounds and bulldogs, race horses and draft horses, and most breeds of fancy pigeon, the differences between adults of different breeds are much greater than the differences between their puppies, foals, and nestlings. Only one exception emerged—the short-faced tumbler pigeon was almost as different from wild pigeons and other breeds when a nestling as it was when fully grown. Darwin draws a very clear lesson from this: since breeders concern themselves only with the form of the adult animal, they select on that basis. Many of the variations that are selected for appear fully only in the adult form, and (by Darwin's second principle) they appear only then in the offspring that inherit them as well. But a few of the variations do appear, or at least are inherited, earlier in life. The result is that most breeds remain relatively similar to each other as young animals, developing their full distinctness only later in life, but a few do display their differences early. (On the other hand, if breeders were to select carefully for traits that appear in younger animals rather than those that appear in adults, we would expect very early differences to accumulate. The young would quickly begin to differ, perhaps becoming as different as the adults of different breeds are today.)

So the analogy with artificial selection supports Darwin's hypothesis that the changes that arise through evolution by natural selection generally occur later in development, because that's when that an organism has to provide for itself. When an organism needs to provide for itself, its variations are subject to selection. But early developmental stages are often over and done before an organism reaches this point. Still, "whenever it comes on, the adaptation of the larva to its conditions of life is just as perfect and as beautiful as in the adult animal" (Darwin 1859, 440). Even in such cases, though, Darwin notes that the embryos and larvae of related species tend to resemble each other more closely than the adults do. For instance Cuvier, in his study of adult barnacles, did not see that they were crustaceans at all. But when we look at their free-swimming larvae, their status as crustaceans is obvious.

This short précis can't do *The Origin* justice. It leaves out too many of Darwin's observations, too many of his insights and his arguments. *The Origin* is a classic of nineteenth century science. Beyond laying out Darwin's theory and the evidence for it, *The Origin* also demonstrated approaches and methods and raised questions and issues for further research that have kept biologists busy ever since. The point of our summary here, aside from giving a general overview of the book, was to give a sense of the richness and detail of Darwin's achievement that would help readers to understand its immense impact on biology.

THE IMPACT OF THE ORIGIN

Darwin's case was very persuasive. It succeeded brilliantly with one of the two points he had argued for: despite some early negative reviews, within 10 years of *The Origin*'s publication most biologists accepted that Darwin was right about evolution and common descent. These ideas became part and parcel of the working biologist's outlook on the world. Intense popular attention also came immediately, with reviews in the press and well-attended public presentations and debates. But popular acceptance came more slowly. In fact, popular acceptance of evolution is still a minority view in the United States, where religious objections (arising mostly from a revival of biblical literalism in the early twentieth century) are still influential. More important for the history of science, though, is the fact that natural selection was not nearly as widely accepted among scientists as the fact of evolution.

There were three main reasons for this. First, Darwin had no settled theory of heredity. The basic facts of heredity—that offspring tend to resemble their parents, and that new, selected traits can be passed down—were not in question. The success of breeding made these points clear whatever the mechanism underlying them. But for scientific purposes, the details of the mechanism matter. For example, if traits of parents were blended in their offspring, as many

Proof in Science

Most philosophers of science in Darwin's day were critical of Darwin's scientific method. Darwin argued for his theory by showing that it organized and explained a wide range of facts in biology, and by carefully examining and answering evidence that seemed to count against it. Influential philosophers including William Whewell and John Herschel objected to this method because, they claimed, it could not provide conclusive evidence that his hypothesis of evolution by natural selection was really true. But their philosophical arguments stood on shaky ground: they all claimed that the evidence for Newton's physics was conclusive. Yet Newton's physics was no more *proven* than Darwin's theory. Despite the efforts of philosophers to argue that Newton's physics could be known to be true because it had been inferred from the phenomena in a way that eliminated any alternatives, Newton's physics was superseded in the early twentieth century. Like evolution, the real success of Newtonian physics was its ability to organize, explain and predict a wide range of physical observations.

Just like evolution by natural selection, Newtonian physics explained and predicted many kinds of observations. But this did not show that Newton's principles were true, or that they would pass every test covering every physical phenomenon. The principles of Newtons' physics imply many things about observations that have never and will never be made. Since we can't know that these observations would also support Newton's principles, we can't know that the principles are true. Success at explaining a wide range of observations combined with the absence of known counter-examples is all that scientists can expect of their theories, no matter what field they work in.

In fact, the word *proof* originally meant "test," as in "the proof of the pudding is in the eating," or "the exception proves the rule" (which does not mean that we somehow prove a rule by finding exceptions to it, but instead that we *test* a rule by observing what happens in exceptional cases where its predictions are surprising). The word *proof* in the more recent and much stronger sense of a conclusive and complete case that leaves no room for doubt is something that we find only in mathematics—where we can specify our starting premises and the rules for reasoning completely and explicitly. But when we propose a theory that's actually about the physical world, the theory is vulnerable to what actually happens when we observe the world. No matter how successful such a theory is, there is no guarantee that it won't fail the next test. Still, we have been very successful at building reliable theories—theories that make correct predictions of novel observations over and over again, despite this vulnerability. Both evolution and classical physics are among the most reliable theories ever developed.

Probability in Evolution

One special feature of evolution by natural selection is that many of its predictions involve probabilities. Natural selection predicts that a population regularly exposed to a poison (say, fruit flies that are controlled with the help of some insecticide) will probably evolve resistance to the poison over time. This is probable but not certain because natural selection treats new traits as arising by chance—being exposed to insecticides does not guarantee that a mutation that helps protect against the insecticide will occur. In fact, it doesn't even make such a mutation more probable. All that exposure to the insecticide does is to ensure that, if and when such a mutation occurs, it will be selected for. Furthermore, probability comes in again because we cannot predict what sort of mutation will occur. Some insects have acquired resistance by developing mutated enzymes that rapidly break down the insecticide. Others have acquired better pumps that remove the poison from their cells more rapidly. And others have altered their behavior, reducing the amount of insecticide that they are exposed to. Any variation that has these effects will tend to spread. But this prediction is also a matter of probabilities, since the details of exactly which insects carry the mutation and what happens in the particular area they inhabit and how many of their offspring inherit the mutation and survive other dangers can't be entirely predicted either.

By contrast, classical physics uses deterministic models, in which fully specified initial conditions for a system allow the prediction (in principle, if not in practice) of its later states, assuming that the system is not interfered with from outside. A general reluctance to accept indeterminacies and probabilities as part of science contributed to the widespread resistance to natural selection as the mechanism of evolution: Herschel famously described natural selection as "the law of higgledy-piggledy." Ludwig Boltzman soon encountered this same reluctance in his work on statistical mechanics, which explained the apparently deterministic theory of thermodynamics by applying statistical probabilities to the mechanical interactions of atoms and molecules.

Vera Causa and Consilience

Darwin was acutely aware of the philosophical objections his theory would face. He carefully designed his argument in *The Origin* to answer those objections as best he could. Philosophy of science in Darwin's England was generally *inductivist*. Figures like Herschel, Whewell, and Mill held that the scientific method included a special kind of *inference*, in which we reason from facts about our observations to a general principle uniting and explaining those facts. For Whewell, thinking through the evidence gathered by observation and considering a range of ways in which it could be explained and unified would lead us to infer the right general idea for the job; Whewell called this process of reasoning *discoverer's induction*. Unlike Mill, who was skeptical about any conclusions science might reach about unobservable things like gravity, Whewell felt that these general ideas could be relied on even when they reach beyond our observations. Darwin, who was also trying to support conclusions that reached beyond any observations we can make, offered an argument that drew on Whewell's ideas, though it did not satisfy Whewell.

To establish a *vera causa* or true cause, Whewell held that we had to combine discoverer's induction with predictive success, successful application to new kinds of phenomena, and increasing coherence. For Whewell the great example of science done right was Newton's physics. Gravity, of course, is itself an invisible force. According to Whewell, Newton's law of gravity was the conclusion of a discoverer's induction, in which a feature or pattern shared by many observations was seen to be a general principle. Subsequently, Newton's law enjoyed all the successes that Whewell demanded of a *vera causa:* it made correct predictions, was successfully applied to a wide range of new phenomena, and grew more and more coherent as these phenomena were connected together and as the mathematical foundations of classical mechanics became better and better understood.

Unlike Newton's theory of universal gravity, the process of natural selection is grounded directly in common sense knowledge of life. Darwin's analogy between natural and artificial selection appeals to straightforward, observable facts—the heritability of traits, the tendency of living things to out-breed their resources and the consequent struggle for existence, the fact that even small differences can have an important effect on survival. Given these familiar facts, natural selection will inevitably occur—but this alone doesn't establish its *importance* to evolution; other factors could have been more influential. To show that natural selection is a *vera causa* of evolution by Whewell's standards, Darwin needed successful predictions, applications to new phenomena, and increasing coherence.

Darwin's responses to the objections in the middle chapters of *The Origin* help with this, by showing that what might superficially seem to be predictive failures are actually not. With this task accomplished, Darwin turned next to apply evolution by natural selection to other biological facts, using it to explain the distribution of life through space and time and the branching patterns of taxonomy and development. These go a long way towards establishing predictive success, successful new applications, and growing coherence, as the details and explanatory scope of natural selection are extended. But much of the work here is done by common descent, not natural selection. Without an account of heredity, the effects of natural selection over time could not be fully tested and confirmed, and natural selection remained an important hypothesis, not an established true cause.

believed, then even advantageous new traits would tend to be washed out over a few generations, especially in large populations. Matings with the rest of the population would spread the new trait thinner and thinner, until it no longer made a significant difference to survival and reproduction. So a wild population would not be changed in any significant way even if very favorable variations occur from time to time. A complete account of natural selection needs a theory of heredity that solves this problem.

Second, many scientists thought that other processes were at work. Some believed that Lamarckian striving would cause new traits to arise in individuals, and their offspring would inherit the traits their ancestors had acquired by hard work. Until experimental work on breeding undermined the idea that acquired characteristics could be inherited, this idea struck many as more convincing than Darwin's view that variation is undirected. At least on the Lamarckian view there would be an intrinsic tendency for variations to be favorable. Two general themes characterized these efforts to find another process responsible for evolution: *directed* variations, that is, an internal tendency either to adaptive variations (Lamarckism), or to variations tending in some particular direction (orthogenesis), and sudden variations, in which new species arise in a single step (saltationism).

Orthogenesis was popular among paleontologists, many of whom claimed to have found general tendencies characterizing fossil lineages. Edward Cope, the discoverer of many famous North American fossils, described sequences of species in the fossil record that seemed to grow larger and larger over time. His reconstruction of the horse lineage followed this pattern, beginning with the relatively small *hyracotherium* (sometimes called eohippus, the dawn horse), and finishing with the modern horse. Cope regarded this sort of trend as evidence both for

The Creativity of Natural Selection

Charles Lyell came to accept evolution, but never accepted natural selection as the main cause of evolution. He appealed to an analogy with Hindu theology, claiming that natural selection could serve as Vishnu, the preserver, and as Shiva, the destroyer, but not as Brahma, the creator. That is, he accepted that natural selection would preserve a well-adapted species and destroy any maladaptive variations. But he never accepted that natural selection could produce new adaptations and new species. This view persists today among a few skeptical biologists and many non-scientific critics.

The biologists in question emphasize that natural selection depends on variation; without new variation to work on, natural selection will stall. Consequently, they argue, we should regard the source of variation as the truly creative force in evolution. However, if variations are undirected, and our evidence continues to show exactly that, then the variations would never lead to any useful new features without the guidance of natural selection. Further, as R. Fisher argued in 1930, the power of natural selection to alter the genes of a species is far greater than that of mutations, which are too rare to make much difference by themselves. Natural selection needs variation to work on. But undirected variation is certainly not creative all by itself. By itself, variation is pointless and powerless to produce any interesting result. Natural selection sifts the variations and builds many copies of the most successful ones, making it possible for new variations to arise some of which will also be selected for. The complete process of variation and selection is creative, even though neither the variations nor natural selection alone is. So Lyell was wrong. Darwin's theory has room for Brahma, alongside Vishnu and Shiva.

orthogenesis and for some kind of Lamarckian pattern of inheritance, arguing that selection of random variations could not explain such long-term trends. Many paleontologists believed such trends were due to innate tendencies, playing the same role in species and lineages of successive species that development plays in individual organisms. They were supposedly independent of selection, and would continue even if their effects were driving the group to extinction. One standard example of this was the Irish Elk, whose massive antlers, it was claimed, led to its extinction as they grew steadily larger and larger. Some even held that species and lineages had a natural lifespan, moving from growth and vigor in their youth to a kind of senility characterized by the emergence of extreme and maladaptive traits, ultimately driving the lineage into extinction.

Third, in the last half of the nineteenth century a major debate arose over the age of the earth. William Thompson, who later became Lord Kelvin, was one of the most famous physicists of the century. In 1861 he published a calculation of the earth's age based on the gradual cooling of a solid sphere. He argued for an outside limit of 400 million or so years, but preferred a figure in the neighborhood of 100 million years. Worse, with new data on the conductivity and heat capacity of rock, Kelvin later argued for a reduced allowance of at most 40 million years. Given Darwin's gradualism, this was not enough time for evolution by natural selection. This problem gave more weight to the demand for new processes that could speed evolution up. Evolution on Kelvin's timescale needed faster processes—larger, more dramatic variations, goal-directed changes, or Lamarckian evolution through the inheritance of acquired traits.

These three problems were not resolved until the twentieth century. Worries about the age of the earth were the first to be answered. The discovery of radioactivity by Becquerel in 1896 was soon followed by evidence from Pierre Curie and Albert Laborde showing that large amounts of energy are released by radioactive materials. Thus, rather than cooling passively as Kelvin had assumed, the earth has a source of replacement heat in the radioactive elements of the crust and core. Ernest Rutherford made the point explicit in a speech before the British Society, and before Kelvin himself, saying "Lord Kelvin had limited the age of the earth, provided no new source (of heat) was discovered. That prophetic utterance refers to what we are now considering tonight, radium!" (Eve 1939, 107) The large amounts of energy released by radioactive elements also undermined Kelvin's worries about the sun's energy, which he had also calculated to be only enough for 100 million years or so at roughly present levels of output.

Better yet, radioactive elements and their isotopes provided natural clocks that could be used to measure the age of some rocks. These clocks are based on the simple observation that these elements all have a measurable half-life, a characteristic time in which half of the sample will spontaneously decay; various techniques can be used to measure the amount of decay that has occurred since certain rocks were formed. Radiological dating finally allowed scientists to assign ages to the various geological periods that had been identified by the stratigraphers of the nineteenth century. Refinements to these techniques continue, but the approximate ages of the main geological periods have been settled since the middle of the last century.

How scientists arrived at a theory of heredity and made it part of a more complete theory of natural selection is a longer story, which we take up in the next chapter. But before we end this chapter, one further early piece of evidence in favor of natural selection is worth noting: Batesian mimicry. In a paper published in 1862, Henry Bates reported observations of some brightly colored butterflies in the Amazon. Strangely, butterflies of two entirely different families looked extremely similar—several species of *Pieridae*, normally plain-looking white butterflies, closely resembled brightly colored species of *Heliconiidae*. The resemblance, in shape of wings and color pattern, was so close that they could not be distinguished except when caught and examined.

Each such species of *Pieridae* appears only in the same region where the similar species of *Heliconiidae* is found. In some cases, a single species of *Pieridae* resembled different species of *Heliconiidae* in different regions; these varieties of the *Pieridae* species often blended into each other at the boundaries of their territories. Bates also observed that the *Heliconiidae* were numerous despite flying so slowly that they should be easy prey for predators. Noticing an unpleasant odor when handling them, he concluded that these butterflies tasted bad.

Bates went on to argue that natural selection was the best explanation for the resemblance between *Pieridae* and *Heliconiidae*. Local predators quickly learned not to eat the bad tasting *Heliconiidae*. So a tasty *Pieridae* that happened to resemble the local *Heliconiidae* even a little might sometimes escape being eaten. A single jump to a high level of resemblance was ruled out because the degree of resemblance found actually varied. The environment alone couldn't lead a single species to resemble different *Heliconiidae* in different regions with similar conditions. But natural selection would strongly favor variations that improved on an initial, very imperfect resemblance, producing resemblances ranging from the crude to almost perfect. Once again, the existence of intermediates argued for Darwin's gradual process of natural selection over sudden jumps.

Time Scales (millions of years) 1961-1975				
			Armstrong	Afanas'yev
		Harland	&	&
	Kulp	*et al.*	Lambert	McDowell	Zykov
	1961	1964	1971	1974	1975
Cenozoic	63	70	65	65	66
Cretaceous	135	136	135	143	132
Jurassic	181	193	200	212	185
Triassic	230	225	240	247	235
Permian	280	280	280	289	280
Carboniferous	345	345	370	367	345
Devonian	405	395	415	416	400
Silurian	425	435	445	446	435
Ordovician	500	500	515	509	490
Cambrian	600	570	590	575	570

Figure 4.1: Table of radiological figures for the ages of the geological periods, cited in Albritton, *The Abyss of Time*. Illustration by Jeff Dixon.

Society and Science, Wilberforce and Huxley

Darwin's illness kept him from taking an active role in the public debates that followed *The Origin*'s publication. The energetic and articulate Thomas H. Huxley took on the role of Darwin's bulldog, speaking out in defense of Darwin's ideas. Other friends and supporters, including Joseph Hooker, also rallied around. The early debate was heavily weighted against Darwin. The view that species were fixed was well entrenched, and evolution had been proposed and rejected before. Religious worries that evolution reduced God's role in the world in general, and about its implications for the special status of human beings in particular, made the debate much more than a dry discussion amongst scientists. Public interest was high, and the initial response of the social elites was very negative. One famous remark on these lines is attributed to the wife of a Bishop: "Let us hope that what Mr. Darwin says is not true. But if it is true, let us hope it will not become generally known." In fact, the shock of evolution's implications about human origins was too much even for most scientists to accept.

The nineteenth century witnessed an increasing professionalization of science. This change brought with it an increasing independence of scientific thought from social and religious constraints. The idea that scientists' work should be guided by evidence and argument alone was becoming more widely accepted, and the social status of science increased as a wider and wider audience came to value our growing understanding of the natural world. But scientists still faced pressure not to directly contradict religious doctrine. Many, including both Darwin and his friend and supporter, Joseph Hooker, censored themselves to avoid public and personal conflicts. Links between universities and religious institutions were still very strong, and publicly endorsing unorthodox opinions could have serious consequences for both reputation and career.

At a meeting of the British Association in Oxford on the thirtieth of June in 1860, the Bishop of Oxford, Samuel Wilberforce, spoke out against Darwin's theory. He denounced it as speculative and contrary to the well-established fixity of species. Late in his speech, having argued that the line separating species cannot be crossed, Wilberforce turned to rhetoric. The subsequent exchange is legendary, both in the sense that it's an often-told tale, and in the sense that there is no reliable record of what was said. But it appears to have gone something like this: in a somewhat indelicate challenge, Bishop Wilberforce asked Darwin's defenders whether it was through his (or someone's) grandfather, or through his *grandmother*, that they claimed he was descended from an ape. Huxley's response was something to the effect that he would rather be descended from an ape than from someone who used the gifts of reason and rhetoric to obscure a scientific issue. (Some reports make Huxley's reply a bit blunter.) By calling attention to Wilberforce's unscientific and impolite appeal to Victorian sentiments about women, Huxley's response put the Bishop on the defensive. Hooker, who had just declared his acceptance of Darwin's views after many years of discussion and correspondence with Darwin, followed up Huxley's rhetorical advantage with a more substantial response to Wilberforce's arguments.

Perhaps more importantly to Darwin's supporters, Huxley's willingness to confront a bishop of the Church of England expressed a claim to equal standing with a man who outranked him socially. Elsewhere, Huxley had emphasized that human ancestry was irrelevant to present human worth, a point with considerable appeal to members of the Victorian middle class, who were anxious to assert their place in a still-hierarchical society.

Later reports make more of the exchange between Huxley and Wilberforce than reports that appeared at the time—this may reflect a growing sense that science had finally established its independence from religious authority. Wilberforce's written review of *The Origin*, on which his speech was based, was a hostile and very skeptical response to evolution. While acknowledging that science must follow where the evidence leads, Wilberforce immediately went on to say, "we shall ask leave to scrutinize carefully every step of the argument ... and demur if at any point of it we are invited to substitute unlimited hypothesis for patient observation, or the spasmodic fluttering flight of fancy for the severe conclusions to which logical accuracy of reasoning has led the way" (Wilberforce 1874, 58–59). Of course, he claims to find these sins against science in Darwin's work.

Although it was fair enough at that time to object to natural selection as, thus far, not proven, and it may just barely have been tenable to object to evolution and common descent in the same way, neither "unlimited hypothesis" nor "fluttering flight of fancy" fairly describes anything in *The Origin*. Darwin had gone to great lengths to acknowledge and answer the objections Wilberforce offered. His answers left room for doubt, but not for Wilberforce's harsh dismissal. Scientists, including Richard Owen, also wrote dismissive reviews. But, like Wilberforce, they suffered from a tendency to oversell their case against Darwin. In fact, Owen seems to deliberately misinterpret Darwin on some points, accusing Darwin of misunderstanding issues he understood perfectly well.

5

HEREDITY AND NATURAL SELECTION

It is easy without any very profound logical analysis to perceive the difference between a succession of favourable deviations from the laws of chance, and on the other hand, the continuous and cumulative action of these laws. It is on the latter that the principle of Natural Selection relies.

Fisher (1930, 40)

INTRODUCTION

Gregor Mendel was an Augustinian monk who had studied at the University of Vienna. He worked as a teacher, mostly of physics, at a monastery in Brno, Austria. But he was also an enthusiastic naturalist, with a special interest in questions about variation in plants. His main interest was how variations are inherited—a subject absolutely central to understanding natural selection. Mendel's work on the subject focused on pea plants: he grew about 28,000 plants in the course of experiments carried out between 1856 and 1863.

As we've seen already, one of the important gaps in Darwin's theory of evolution was the lack of a theory of inheritance. Mendel's work could have filled that gap—sadly, it was published in an obscure journal and didn't receive the attention it merited. Mendel's tendency to underemphasize the more theoretical conclusions he had reached may also have prevented the few who did read his work from understanding its significance. Only at the end of the nineteenth century, when his principles of heredity were rediscovered by Hugo de Vries, was Mendel's accomplishment finally recognized.

But the rediscovery of Mendel's ideas was not enough. Their incorporation into the theory of natural selection was delayed by rivalry and misunderstanding. The proponents of the new genetics saw it as an alternative to natural selection, while the so-called biometricians, defenders of gradual natural selection, regarded the discrete traits of the geneticists as aberrant and (more importantly) irrelevant to understanding the continuously varying traits that

Figure 5.1: Gregor Mendel, discoverer of particulate inheritance. Courtesy National Library of Medicine.

they found in wild populations. In this chapter we'll learn about Mendel's ideas, the early days of the science of genetics, and the early geneticists' criticisms of natural selection. Then we'll see how these disputes were resolved when Mendelian genetics was combined with a statistical treatment of populations, their gene pools and natural selection to create the *modern synthetic theory of evolution*.

MENDEL'S ELEMENTS

Mendel's work on peas examined how some established variations in pea plants were passed along when the varieties were cross-pollinated. Some of his plants produced smooth peas while others produced wrinkled ones; some of the peas were yellow and others green. In all, Mendel examined seven traits that appeared in different strains of peas. Mendel cross-pollinated smooth, yellow varieties with wrinkled, green varieties. We label this parental generation P. The next generation of plants, conventionally labeled F_1 (the first filial generation), all produced smooth yellow peas, just like those produced by half the plants in P. It was as if the wrinkled and green traits had completely disappeared in F_1. This result was already a substantial blow against blending accounts of inheritance. It seemed, at least in some cases, as though one trait completely *dominated* another. But when Mendel bred this hybrid generation together to get a third generation, F_2, wrinkled peas re-appeared in about a quarter of the plants; green peas also reappeared in about quarter of them. What had happened?

Mendel's explanation was simple, but it had immense implications for the nature of inheritance. He proposed that each parent plant contributes one particle or element corresponding to each trait. Each plant inherits two such elements, one from each parent. But the effects of these elements depend on which trait is dominant and which is recessive: a plant with one dominant element and one recessive element will appear indistinguishable from a plant with two of the dominant elements.

The P generation was made up of pure varieties (strains) of peas; one consistently produced smooth peas and yellow pods, and the other wrinkled peas in green pods. Therefore the smooth yellow plant must have carried two elements for the smooth trait, and two elements for the yellow trait; similarly, the wrinkled green plants must have carried two elements for the wrinkled trait and two for the green trait. As a result, the F_1 generation received one each of the two types of elements: one smooth element and one yellow element from the smooth, yellow parent, and one wrinkled element and one green element from the wrinkled, green parent. Since the elements that produce

Parent Elements Received	From P2: S (prob= 1/2)	From P2: s (prob= 1/2)
From P1: S (prob= 1/2)	S,S (prob= 1/4)	S,s (prob= 1/4)
From P1: s (prob= 1/2)	s,S (prob= 1/4)	s,s (prob= 1/4)

Figure 5.2: Probabilities for smooth and wrinkled elements received by f2 offspring. Illustration by Jeff Dixon.

Parent Elements Received	From P2: G (prob= 1/2)	From P2: g (prob= 1/2)
From P1: G (prob= 1/2)	G,G (prob= 1/4)	G,g (prob= 1/4)
From P1: g (prob= 1/2)	g,G (prob= 1/4)	g,g (prob= 1/4)

Figure 5.3: Probabilities for yellow and green elements received by f2 offspring. Illustration by Jeff Dixon.

the yellow and round traits are dominant, all these hybrid plants would produce smooth, yellow peas, just like the smooth yellow plants in P. But the green, wrinkled elements are still there in the F_1 plants, waiting to reveal their presence in the next generation.

Assuming that it was equally probable for a plant to inherit either element from each parent, the F_2 generation would be divided about equally between plants that get the smooth element from both F_2 parents, those that get the smooth element from the first and the wrinkled element from the second, those that get the wrinkled element from the first and the smooth element from the second, and those that get the wrinkled element from the first and the wrinkled element from the second—in the table below, we use upper case for dominant and lower case for recessive, and label the smooth element S and the wrinkled element s.

Of course smooth elements are still dominant, so three-quarters of the plants in F_2 will produce smooth peas and only a quarter will produce wrinkled peas. The same relation holds for the color elements—each plant in F_2 has an equal chance of receiving either the green or the yellow element from each parent. So the table for the color elements looks like Figure 5.3

Since yellow is the dominant element here, three-quarters of the plants in f_2 will produce yellow peas, and a quarter (those that inherit the G element from both parents) will produce green peas. These traits are inherited independently of each other, so we can give a similar table for both traits:

Parent Elements Received	From P2: S,G (prob= 1/4)	From P2: S,g (prob= 1/4)	From P2: s,G (prob= 1/4)	From P2: s,g (prob= 1/4)
From P1: S,G (prob= 1/4)	S,S,G,G (prob= 1/16)	S,S,G,g (prob= 1/16)	S,s,G,G (prob= 1/16)	S,s,G,g (prob= 1/16)
From P1: S,g (prob= 1/4)	S,S,g,G (prob= 1/16)	S,S, g,g (prob= 1/16)	S,s,g,G (prob= 1/16)	S,s,g,g (prob= 1/16)
From P1: s,G (prob= 1/4)	s,S,G,G (prob= 1/16)	s,S,G,g (prob= 1/16)	s,s,G,G (prob= 1/16)	s,s,G,g (prob= 1/16)
From P1: s,g (prob= 1/4)	s,S,g,G (prob= 1/16)	S,S,g,g (prob= 1/16)	S,s,g,G (prob= 1/16)	s,s,g,g (prob= 1/16)

Figure 5.4: Probabilities for both traits combined in f2 offspring. Illustration by Jeff Dixon.

The plants represented in the bottom right corner of this table have wrinkled green peas; breeding from them produces only wrinkled green peas, just like the strain that was used as one parent in the P generation.

These results show, at least for these particular traits, that blending does not occur. Instead of passing along a mixture of each parent's traits, these elements seem to be passed along unchanged, ready to mix with other elements in later generations. This provides an answer to one of the most difficult early objections to natural selection, the *blending* objection.

As we've already noted, Mendel's work went unnoticed. He did send a copy of his 1866 paper to Darwin, who was overwhelmed with correspondence at the time. Darwin never opened the letter. Darwin himself, hoping to fill the gap in his theory, proposed an account of heredity in his *Variation of Animals and Plants under Domestication,* published in 1868. The theory, which Darwin called *pangenesis,* proposed that particles were drawn from all parts of the body and combined in the germ cells (eggs and sperm), which join together to produce offspring. This picture of heredity fit the facts as known. Parental traits could be passed along to their offspring; it allowed for the inheritance of acquired characteristics also, since these particles would transmit changes in the parent's bodies to the germ cells in their reproductive organs. But Darwin supposed that external influences on the parents' bodies and sexual organs would be the cause of most variations.

AFTER MENDEL

Darwin's cousin, Francis Galton, soon produced experimental results that put Darwin's proposal in doubt. Galton performed blood transfusions on pure-bred rabbits, using blood from differently colored breeds. He then bred the rabbits, to see if coat colors from the rabbits used as blood donors would show up in the offspring. But no changes in coat colors were observed—though, on Darwin's pangenesis theory, there ought to have been particles in the blood carrying the coat-color trait of the donor. Galton concluded that inheritance is transmitted not just from parents, but from all ancestors; whatever hereditary material we possess, on his account, is inherited from our ancestors and passed down unaltered. It is not manufactured by our bodies, but simply passed on through them.

Variation, from Galton's point of view, was simply a range of traits appearing in a population; heredity then transmits the range of traits present in the ancestors on to descendent populations. This focus on inheritance at the population level was a useful point of view for many biologists, but it was a turn in the opposite direction that led to new insights on the mechanism of heredity: microscopic studies of cell division showed that new cells were formed from previous cells, and that the new cell nuclei were formed from the nucleus of the original cell; studies of egg and sperm cells showed that the nuclei of these cells united at fertilization. The hereditary material, it seemed, was stored in the nucleus.

Further work added detail and conviction. New observations of the nuclei of dividing cells noted the presence, in the nucleus, of material that was strongly dyed by newly developed synthetic dyes. The material was named *chromatin,* for its strong coloration by these dyes. As a cell prepared to divide, the chromatin was drawn together into a number of rod-shaped units called *chromosomes.* The chromosomes were duplicated by dividing lengthwise, and the copies of the divided chromosomes were drawn to opposite poles of the cell. The cell then divided across the middle, and the chromosomes gathered together to form the nuclei of the new cells.

Since germ cells combined their nuclei to produce a new individual, it seemed they must each contribute half the necessary chromosomes. August Weisman concluded that a cell-division without duplication of chromosomes must occur in the formation of these cells. His proposal was confirmed when meiosis was observed in the late 1880's. Slightly more complicated than Weisman's initial proposal, it produced the same result by first copying the chromosomes to produce a double set, and then dividing twice to produce four germ cells.

The cells that produce germ cells establish a separate lineage early on in development. As a result, there seemed to be no way for information from the rest of the body's cells to influence what is passed on through the germ cells: Galton's vision of ancestral inheritance was borne out, and the inheritance of acquired traits suffered another blow. In fact, without any way for the organism's efforts in life to alter the hereditary material carried in the germ cells, directed variation seemed to be ruled out in general. New variations, it seems, can not be triggered by the organism's needs; they arise at random in that sense—though of course not just any variation can occur. Each variation that arises occurs in an already existing type of organism and must interact with the development and structure of that type.

Many at the time still found the idea of sudden jumps in evolution attractive. William Bateson's *Materials for the Study of Variation* appeared in 1894. In his book, Bateson argued that variation, not selection, is the key to evolution. Distinguishing between *continuous variation,* the normal range of differences between individual members of a population, and *discontinuous variation,* variation that falls dramatically outside the range characteristic of the population, Bateson emphasized a parallel between species as we find them, with clear gaps between them, and the process that produces the species, which he saw as making a jump across those gaps. Bateson also pointed out that many traits can't be treated in terms of continuous variation—for example, the number of needles in a bundle on a conifer tree must be a whole number, such as one, two, or three. It can't be 3.15, or any other fractional number. A particulate theory of heredity was crucial to this story, since these discontinuous variations had to persist in the population without dilution or blending, if a new species was to result.

At the same time, a botanist named Hugo de Vries was also studying heredity. Like Mendel, he thought in terms of elements or particles underlying individual traits; each, he proposed, could be inherited or altered independently of the others. This led him to a series of breeding experiments much like Mendel's, and the patterns Mendel had discovered years before began to reappear in de Vries' plants. A careful review of the literature before publishing his findings led him to Mendel's work, rediscovered at last.

To de Vries and Bateson, Mendel's talk of elements was transparent—these were the units of heredity, each specifying a different trait, and independently distributed to offspring by their parents. The particles received from each parent operated together to determine the features of the offspring, but were passed on separately and independently to the next generation. New terms were invented to describe the components of the theory: *allele* became the term for the individual particles connected to a particular trait. A fertilized egg was called a *zygote*. When speaking of a particular trait, it was a *homozygote* if it had two copies of the same allele (as in the *P* generation of pure varieties in Mendel's experiments), and a *heterozygote* if it had one copy each of two different alleles. Careful work revealed the patterns governing more complicated cases, showing that these could also be understood according to the basic rules Mendel had developed.

But what were the implications for evolution? De Vries and Bateson agreed on the importance of Bateson's discontinuous variations. The new genetics showed that blending or dilution would not occur. New forms of alleles were passed on without alteration, at least to some offspring. De Vries studied a particularly striking case of discontinuous variation in the evening primrose—a variation that bred true, and was accompanied by further, striking variations in other plants' offspring. It seemed to de Vries that the species was breaking up and dividing before his eyes. The word *mutation* was coined by de Vries to describe this process. Though he could not find other examples of this strange behavior, de Vries simply concluded that the other species he examined were not undergoing mutations at this time. If a single mutation involved enough change to produce a new species, then the idea of natural selection slowly building up differences over time before a new species emerged was unnecessary. Natural selection could act as a kind of gatekeeper, eliminating any new species that failed to survive

Figure 5.5: Hugo de Vries rediscovered Mendel's ideas. From Samuel Christian Schmucker, *The Meaning of Evolution*. New York: The Macmillan Co, 1913. Courtesy the University of Calgary Library.

and reproduce in competition with its ancestors. In this way, natural selection would remain both preserver (of a successful species) and destroyer (of unsuccessful ones). But natural selection would no longer be regarded as the creator of new species—de Vries assigned that role to his mutations.

THE KEY TO EVOLUTION

The struggle was on: defenders of natural selection were now opposed, on one side, by Lamarckians who continued to advocate directed variation, and on the other, by the new geneticists, who saw mutation as the creative force in evolution. Some of them resisted Mendelian genetics, arguing that not all traits follow the Mendelian rules and emphasizing the evidence for continuous variation in real populations. But further work by geneticists showed that continuous variation could be produced by Mendelian genes. Even plants with identical genes vary depending on their environments. Working with genetically pure strains of beans, Wilhelm Johannsen selected extremes from a range of sizes and planted them. The resulting offspring had the same range of sizes as the original population, showing that this kind of variation is not heritable. A wider but still continuous range of variation appeared in mixed populations, but there it could be selected for—each strain had its own characteristic range of sizes. Variations due to the environment had smoothed out the discontinuous distribution of genes, turning it into the continuous distribution that naturalists had so often observed.

This work clarified the distinction between what Johannsen called *genotype* and *phenotype*. An organism's genotype is the set of heritable traits encoded in its genes. But its phenotype is its actual physical form, the product of those genes and the environment in which it developed. Observable variation arises both from the environment and from the genotype, often producing continuous variation in the population despite the fact that the genetic differences are discrete and not continuous. A wide range of continuous variation is especially likely when a number of different alleles contribute to a particular trait like the size of beans. There was no need to posit two distinct forms of variation anymore—discrete Mendelian genes together with the influence of environmental differences could account for both the sharp and dramatic differences emphasized by the early geneticists and the continuous variation that naturalists had observed.

Thomas Morgan was the first geneticist to work extensively with the fruit fly *Drosophila* as a laboratory animal. The ease of keeping and feeding these flies together with their rapid breeding cycle made them ideal for genetic studies. His work extended Johannsen's, demonstrating that dramatic changes like those proposed by de Vries were not the only kind that could occur. A mutation in the modern sense is a change in some gene (allele); a mutation's effects can range from being quite small (in fact, some mutations have no effect on the phenotype of an organism) to very dramatic, depending on just what that altered gene does and how. Morgan's careful study of his flies, generation after generation, showed that new mutations are arising all the time, many of them with

Figure 5.6: Thomas Morgan in the fly room at Caltech. Courtesy of the Archives, California Institute of Technology.

only minor effects. Following how these mutations were passed on to later generations allowed Morgan and his team to refine and extend Mendel's ideas into a rich and powerful theory of inheritance.

This reconciliation of Mendelian inheritance and continuous variation in natural populations opened the possibility of a reconciliation between natural selection and the geneticists. As Darwin had always insisted, given a plentiful supply of small variations, natural selection could gradually construct complex and intricate adaptations. The steady stream of small mutations Morgan and other geneticists had found in their studies looked exactly like what Darwin's theory needed. Moreover, Johannsen's evidence that environmentally induced variations were not heritable reinforced experiments showing that many other kinds of acquired traits were not inherited either. Lamarckian views were rapidly becoming less and less tenable.

Edwin Goodrich developed this argument in his 1912 book, *The Evolution of Living Organisms*. To show that small variations could be significant enough to be selected for, Goodrich appealed to the peppered moth, *Biston betularia*. Once a light-colored moth, a dark form had become common in northern England. Goodrich suggested the dark form had spread because it was better camouflaged in the sooty environment around the industrial cities of the north. He also cited observations of sparrows collected from the ground after a snowstorm in Rhode Island: measurements of their overall length, wingspread, length of head and beak, length of legbones and breastbone, and width of skull by zoologist Hermon Bumpus showed that the birds that recovered were on average slightly larger than the birds that died.

Without a more detailed integration of natural selection and Mendelian genetics, however, the debate was a stalemate. A more general and complete account was needed. In the meanwhile, many geneticists went on emphasizing discontinuity and the role of mutation in evolution, while naturalists emphasized continuous variation and natural selection. The modern synthetic theory of evolution finally emerged only when a new generation of theoreticians developed a mathematical account of how selection would alter the genetic composition of a population over time did.

THE MODERN SYNTHESIS

The first step was an account of *equilibrium:* when no selection occurs, that is, when the alleles are equally fit, a stable genetic balance tends to arise. Consider a pair of alleles, A and a, distributed in a population. Let p be the number of instances of allele A divided by twice the population (since each member carries two alleles); similarly, q is the number of instances of a divided by twice the population. Then we know that $p + q = 1$, since every allele in a population is either A or a. In 1908 G. H. Hardy showed (assuming mating is random) that the distribution of the genotypes AA, Aa, and aa in the next generation will be in the ratio $p^2 : 2pq : q^2$. This distribution will not change in subsequent generations unless one allele out-reproduces the other, whether by chance or by selection. A treatment of how this equilibrium would change under selection was next, and the results were encouraging. Even if one allele out-reproduces the other by a small percentage in each generation, the allele would spread through the population over time. Gradual evolution through the replacement of alleles over time began to look like a good prospect. But a general theory would require a more thorough mathematical treatment.

Ronald Fisher was a statistician with a serious interest in evolution. He began to work on statistical models of genes and selection in 1918, and published a book titled *The Genetical Theory of Natural Selection* in 1930. His calculations showed that most change in gene frequencies is due to selection. Mere chance can make a difference only in small populations or over long periods of time. Fisher showed that, if n is the size of the population, the average time required for chance to decrease variance in the population (a measure of the variability of genotypes) by half is $1.4n$ generations. So, in a population of a million with one year per generation, this amount of change would take an average of one million four hundred thousand years. Mutations alone, on the other hand, are too rare to play an important role unless selection favors them and spreads them through the population. Most genes in the fruit fly *Drosophila* experience mutations in fewer than 1 in every 100,000. So it would take an extremely long time for a significant change in the frequency of some gene to arise simply by mutation. And if it were even slightly disadvantageous, selection would ensure that it never becomes common in the population. But a gene that is rare at one time can quite quickly become very common if it is selected for—that is, if the individuals that carry it consistently tend to survive and reproduce more successfully than the average individual in the population. A gene that provides a mere 1 percent advantage would spread from 26.9 percent to 73.1 percent of the population in 200 generations. On the other hand, a mutation which produces a 1 percent disadvantage cannot spread to affect more than 1 in 1,000 members of the population, even if the same mutation occurred repeatedly at the rate of 1 in 100,000.

Fisher's book presents a statistical theory of natural selection based on Mendelian genetics. His argument from genetics to the importance of natural selection is straightforward: once blending inheritance is given up, variation can persist in the population for far longer, since each allele is passed on undiluted to any descendants who inherit it. Therefore the rate of mutations required to explain present variance is much lower, given Mendelian genetics, than it would have to be on the blending theory. But unless the rate of mutations is extremely high, selection is a much more powerful cause of change in the population than mutation. Further, Fisher argued that the idea that mutations might be directed towards some end, in a Lamarckian or orthogenetic way, is also rendered insignificant by Mendelian genetics. Since mutations are so rare, and selection so powerful at choosing which mutations will spread and which will be eliminated, any directional tendency in the mutations will be dominated by natural selection.

The second chapter of Fisher's book lays the groundwork for an account of natural selection and genetic change over time. Beginning with an elegant equation expressing average mortality and reproduction in a population, Fisher adds in the effects of genes on traits, and of traits on rates of survival and reproduction to arrive at a mathematical account of natural selection.

This turned the tables on de Vries and Bateman: despite their attempt to replace natural selection with mutation, their own results in genetics actually implied that natural selection is the most powerful process causing genetic change in populations. J. B. S. Haldane, a brilliant and mathematically adept biologist, arrived at similar results during the 1920s. Along the way, he applied his mathematics to the case of *Biston betularia,* the peppered moth. The dark form was due to one dominant gene, as breeding experiments had shown. Haldane calculated that for the dark gene to spread as rapidly has it had, it must have had a 30 percent selective advantage over the lighter form. With selection pressures that high, natural selection operates very quickly indeed.

The third figure involved in the development of population genetics was the American biologist Sewall Wright. A brilliant student with great mathematical talent, Wright wrote his doctoral thesis on the inheritance of coat colors in mammals at Harvard. This work, carried out by Sewall and by his adviser William Castle, helped to establish how different genes can interact. Some genes alter how other genes express themselves: instead of coding for a single specific trait, they interact with other genes and the environment to produce a particular phenotype.

Wright and Castle's lab work supported Fisher's contention that genetic variety could persist in natural populations because one gene's consequences for an organism's fitness depend on the other genes it is associated with. In some cases, this interaction makes heterozygotes more fit than homozygotes, leading selection to favor a balance between two alleles and thereby keeping both forms of the gene in the population. The classic example of this is the human sickle-cell anemia gene. While homozygotes for the sickle-cell gene have severe, usually fatal anemia, homozygotes for the normal gene are more vulnerable to malaria than the heterozygotes who have both a normal gene for hemoglobin and a sickle-cell gene. The result is that in areas where malaria is common there is selection for the heterozygotes. This keeps both genes in the population, at levels that reflect a balance between the advantages of heterozygotes and the comparative disadvantages of the two homozygote forms. In other cases, two genes can interact to create advantages for one allele in some genetic combinations and for another allele in others. The result, again, is that selection tends to keep both alleles in the population at some balancing level.

Another interesting kind of case arises when the advantage of having an allele depends on its frequency: an allele that is very advantageous when it's rare enough can become disadvantageous when it is too frequent. So-called cheaters among cleaning wrasses are a nice example of this sort of trait. Cleaning wrasses are small fish that make their living by cleaning parasites and dead tissue from other fish—including their mouths. It sounds like a dangerous lifestyle, but the larger fish cooperate with the wrasses, and the arrangement normally benefits both. The wrasses receive easy meals while the other fish wind up with cleaner, healthier mouths. But some wrasses cheat, taking bites out of their clients. (Of course they do not cheat predatory fish, who might respond by eating them!) So long as the cheaters are rare enough, they do very well—they wind up with extra food at the expense of their cheated clients, who generally startle and swim away. But if they become too common, the wrasses will lose clients, and the cheaters (along with other cleaning wrasses) will lose their easy lifestyle. So a cheater gene would very probably have a frequency-dependent effect on fitness. In fact, wrasses have evolved a more complex strategy, providing slow service and often cheating local fish that don't travel far enough to shift their business to another cleaning station, but providing much quicker service and very rarely cheating their more mobile customers.

After receiving his PhD, Wright took a position in the U.S. Department of Agriculture, where he began work on breeding and selection. Wright recognized that inbreeding a line of plants or animals could identify interesting and desirable traits—because these inbred lines possess a very limited selection of genes from the general population, particular combinations involving those genes will occur much more frequently in the inbred lines. Special traits that emerge from these combinations can become fixed in these lines, and then crossbred with other lines carrying other desirable traits. This technique allowed Wright to quickly establish superior breeds with combinations of traits that would have required decades of ordinary breeding.

Wright's ideas about breeding led him to a different view of when evolution by natural selection would proceed quickest. Haldane and Fisher believed large populations would evolve most rapidly, because they would contain a wider range of different alleles and would generate larger numbers of new variations. But Wright proposed that a large population divided up into largely isolated sub-populations would evolve quickest. Each sub-population would be like one of his inbred lines; in them, gene combinations that were very rare in the wider population could become common. Some of these combinations

Lysenkoism

The group Dobzhansky had trained with in Russia was at the forefront of genetics and natural selection in the late 1920s. But things there changed dramatically, shortly after Dobzhansky left for the United States. Neither Soviet biology nor Soviet agriculture ever recovered from the damage. In 1927 Trofim Lysenko, a plant breeder with no scientific training, was proclaimed a peasant genius. It was said that he could fertilize fields without fertilizers or minerals and grow peas in Azerbaijan's winter. Lysenko gained immense political clout by a close association with Stalin, though his ideas were bizarre and wildly indefensible. Lysenko rejected the idea that genes limited when and how a plant could grow—this idea, he claimed, was anti-Marxist, because it assumed a kind of essence that limited the nature of life, where true Marxism held that life was infinitely perfectible. Placed in the right environment, seeds of wheat could even grow to be rye. Lysenko held that Mendelian genetics and natural selection were bourgeois theories based on a competitive, capitalist view of nature. A socialist biology would emphasize cooperation instead: seeds planted together would sacrifice themselves to help one among them grow to new heights of health and productivity. Tests and evidence were not needed—Stalin had already proclaimed the preeminence of practice over theory in 1929. The practical success of Soviet agriculture would be all the proof that was needed. Though project after project failed, triumph was always just around the corner. In the meanwhile, Soviet biology decayed as Lysenko's opponents were fired, imprisoned, and even killed.

For all of its present status and success, science is a fragile thing. Galileo spent the last years of his life under house arrest because, in his *Dialogue on the Two Chief World Systems*, he endorsed the sun-centered model of the solar system as true. He had already been warned by the Inquisition to treat it only as a hypothesis that "saved the phenomena," that is, that fit the observed positions of the planets and the sun. His rebellion against the Church's authority over astronomy carried a heavy

would produce successful new traits, which would then spread rapidly through the main population when their isolation came to an end. Unlike large populations, these small sub-populations would also experience significant genetic drift. This drift would extend the range of genetic space explored by the population and allow it to reach adaptive peaks that could not be reached simply by selection.

Despite these disputes, however, Fisher, Haldane, and Wright all agreed that natural selection, as modeled by their theories of population genetics, was the key process driving evolution. With the basic mathematics of natural selection settled, biologists moved on to applications. Breeding experiments and fieldwork soon produced strong evidence for the new theory—in some cases, despite the initially skeptical views of the researchers. The results were applied to a growing range of examples.

THE SYNTHESIS APPLIED

Dobzhansky

Theodosius Dobzhansky was a Russian biologist, trained in Kiev and Leningrad. After beginning with work on variation in ladybird beetles, Dobzhansky found their genetics too complicated to work out. He then switched to Morgan's fruit flies, *Drosophila melanogaster*. In 1927 he received an international scholarship and emigrated to the United States. There he joined Thomas Morgan at Columbia University to work in what was called the fly room. He moved to the California Institute of Technology with Morgan in 1930. Dobzhansky turned to studying a species of wild fruit flies, *Drosophila pseudoobscura*, using the techniques of fruit-fly genetics developed in Morgan's labs. His microscopic studies of markers in fruit fly chromosomes showed that flies from neighboring regions were genetically similar, while more distant populations of the same species showed greater differences. In his 1937 book *Genetics and the Origin of Species*, Dobzhansky proposed a new definition of evolution based on population genetics: evolution is a change in the frequency of an allele within a gene pool. This definition sharply focused how the new theory of heredity and the statistical approach to natural selection changed our understanding of evolution. Evolution is a change in the hereditary make-up of a population—in terms of population genetics, this must be a *statistical* change in the population's *gene pool*.

Dobzhansky's account of the origin of new species turned on the accumulation of genetic variation to the point at which two populations lose their ability to interbreed. A population that becomes isolated is genetically different from the main population from the beginning: because it's only a small sample of the main population, it's very unlikely to match the frequencies of every allele in the main population. Even without selection acting on it, a small population will experience genetic drift, that is, random changes in the frequencies of different alleles over time. Also, of course, any new mutations will remain confined to the isolated group, even if they are selected for. After enough time, the two populations, genetically different from the beginning, finally become unable or unwilling to interbreed. At that point, a new species has emerged.

Mayr

Ernst Mayr was studying the birds of New Guinea when Dobzhansky's book appeared in 1937. When Mayr began his

price. Nikolai Vavilov, the gifted Russian botanist and geneticist who first identified where many cultivated plants originated, was imprisoned in 1940 and died in 1943 of malnutrition because of his opposition to Lysenko. When powerful institutions set themselves against scientific inquiry, it can become impossible for science to progress. Genetics in the Soviet Union still has not fully recovered from Lysenko's suppression.

Barriers can arise within science, too—but scientific institutions are designed to give new ideas a fair hearing. Journals rely on experts in each field to evaluate papers that are submitted to them. To keep reputation from influencing who gets published, most journals use blind refereeing, in which papers submitted for publication are evaluated by experts who don't know who the author(s) are. The culture of science is also important. Scientists are ordinary people, with ordinary ambitions and hopes; as individuals they are no more objective or honest than the members of other groups. However, like everyone else, scientists care about their reputations, and especially about their reputations among their peers. In science, reputation depends on scientific accomplishments: new observations and discoveries, theoretical innovations and significant *and* reliable experimental results are what count. The standards by which these accomplishments are judged are taught from early on. Detailed, well-documented observations, accurate calculations, and careful arguments—arguments that deal fairly with contrary evidence and objections—are central values in the world of science.

Science depends on the freedom of scientists to pursue their research independently, to report their results openly, to criticize evidence and arguments for accepted views, to question what is widely accepted and defend what is widely questioned. So long as scientific institutions support these activities, and scientists and scientific institutions are not pressured to produce conclusions that favor powerful political and economic interests, our scientific understanding of the world will continue to improve, as new evidence and new ideas emerge over time.

Figure 5.7: Ernst Mayr first proposed the biological species concept. University of Konstanz. Meyer A (2005) On the Importance of Being Ernst Mayr. PLoS Biol 3(5): e152 doi:10.1371/journal.pbio.0030152.

field work he was a Lamarckian, but during the 1930s he became persuaded that the new theory of natural selection was a better way to explain what he saw in the field. Mayr's studies led him to a new answer to a question that his work as a taxonomist continually confronted him with: how to draw the line between one species and another. Darwin had pointed out in *The Origin* that taxonomists often disagreed about this, and things were no easier in the 1930s. Mayr's work on birds of paradise and kingfishers revealed many local variations. Where such groups shared territory and still remained distinct, they were clearly different species. But when their territories didn't overlap it was hard to decide whether they were different species or not. Furthermore, the idea of evolution raises fundamental questions about just what species are.

Aristotle had a straightforward answer to this question: a species is a group of organisms that share the same form or essence. But there is no fixed form or essence in evolution. At any given time, the population is made up of individuals, each with its own genotype. The population as a whole includes a range of genotypes, with different frequencies for different alleles. These frequencies are always changing—that is, as Dobzhansky defined it, evolution is always going on. Over time these changes add up, and one species slowly becomes another or divides into two. But where to draw the line between one and the other seems arbitrary: there is no single, sudden step that marks the change.

Mayr offered a straightforward answer to this puzzle called the *biological species concept.* In his view, two populations count as one and the same species when they would interbreed if they occupied the same territory. This criterion focuses on a fact of real biological importance: by interbreeding, organisms come to share a single gene pool, exchanging and recombining their genes over generations. So for Mayr, species are groups that would share a gene pool *if* they were not spatially or temporally isolated from each other.

Even this very straightforward idea is hard to apply in some cases. *Ring species* are populations spread out in a ring. In one region of the ring they include two overlapping groups that do not interbreed. But as we move out around the ring from this region, each group interbreeds with its neighbors, and those neighbors with their neighbors all the way around. So we seem to have two species by the biological species criterion, but there is no place to draw the line between them. They are linked by a series of interbreeding populations. In time a new gene could spread around the ring from one to the other. In a tenuous and indirect way, they still share the same gene pool.

Although ring species are unusual, a similar problem arises over time for every species. Consider a long period of time—so long that the descendents of an ancestral population have changed enough that the two would not interbreed, even if they overlapped in time and space. By the biological species concept, these populations belong to distinct species. But one is descended from the other, so, like a ring species, they are linked by populations all of which would interbreed. And like a ring species, there is no special point at which a clear line from one species to another is crossed.

This doesn't show that the biological species concept is a mistake—what it shows is that being of the same species, as Mayr defined it, is not *transitive*. That is, if *A* is the same species as *B*, and *B* is the same species as *C*, it doesn't follow that *A* is the same species as *C*. *A* and *B* may be similar enough that they would interbreed, and the same may hold of *B* and *C*, while *A* and *C* are too different and would not interbreed.

This is a logical fact about species that evolve gradually. Some differences between populations are too small to constitute a difference in species. But the differences that do constitute a difference in species are just more and greater differences of the very same kind. So to pick out a single species using Mayr's criterion, we need to begin with a particular population. That population gives us a touchstone for comparison—other populations elsewhere in time and space either would interbreed with it if the two shared territory, or they would not. This means that how we divide the world up into species will depend on what populations we start with. We've already emphasized that taxonomy at higher ranks is focused on the present, because it divides past organisms into groups that reflect the major divisions between today's forms of life. So it seems natural, as well as convenient, to continue this policy at the level of species and treat present populations as starting points for working out where species begin and end.

This doesn't settle the issue of ring species, though. Drawing lines in these will require choosing a population at a particular place as well as at the present moment. For convenience we might try starting with one of the two populations that overlap but don't interbreed. In cases where we have a line instead of a closed loop, we could start with the population at one end of the line. Then we can pick out a species that includes all the organisms that, if they overlapped that population in space, would interbreed with it. Another species would begin just on the other side of this divide and continue until we reach another such barrier to reproduction, and so on. The point is not to capture some objective fact about species here—given the facts and Mayr's notion of species, there can be no uniquely correct system separating all organisms into different species. The point is to find a systematic way of drawing lines that fits all the standard cases and treats the difficult cases sensibly, something which this proposal seems to do.

But there are other problems as well. The resulting lines will be fuzzy, as Darwin had already emphasized. Sometimes males from population *A* will mate successfully with females from population *B*, even though males from *B* will not mate successfully with females from *A*. And in general, since populations include a varied collection of individuals, whether or not population *A* will interbreed with population *B* is really a matter of degree. Sometimes the average fertility of matings between two populations will be lower than for matings within each population, but still well above zero. This raises the question, how much lower would it have to be before we regard them as different species? Lastly, sometimes fertility isn't reduced at all, but hybrid offspring are unable to produce successful mating displays for either group. So gene pools can be separate despite complete interfertility.

These issues of vagueness don't undermine Mayr's idea. They only emphasize the fact that without Aristotle's heritable, fixed essences, that is, features that must be shared by all and only the members of a species, we cannot draw clear, universal lines between all species. The resulting divisions depend on our chosen starting points, and their exact location in the range of populations across space and time will depend on just how strict our criteria for successful interbreeding are.

Perhaps the most difficult problem for the biological species concept is that it doesn't seem to apply to asexual organisms at all. Consider bacteria, for example: they reproduce by simple cell division. There seems to be no gene pool at all in such cases—each individual, no matter how similar it may be to some other individuals, has its descendants exclusively to itself, sharing them with no other individual. So no new genes that arise in one bacterium are ever shared with another bacterium's descendants in the process of reproduction. But biologists still recognize species of bacteria like *Escherichia coli,* based on close similarities. Here we may have a different notion of species altogether—a purely similarity-based notion, instead of one based on interbreeding and a shared gene pool.

The separation of lineages in asexual organisms is complicated by the fact that genes can be shared in other ways than by mating and reproduction. Bacteria carry some of their genes in *plasmids*, rings of DNA that can be exchanged with other bacteria. This kind of gene-sharing is one reason why resistance to antibiotics can spread so quickly in bacteria exposed to them. More dramatically, viral genes can become part of an organism's genome and be passed down to their offspring. In fact, some human genes are homologous to known virus genes. So there is a sense in which all life shares a single gene pool, given enough time. Of course there is still very good reason for treating the gene pools of most species as more or less isolated. Allowing for a few exceptions like plasmids, the amount of such exchange is very limited compared to what is inherited from parent(s). However, this may not have always been true. Early in the development of life, the exchange of genes between different lines of descent may have been very important. Instead of descending along a single trunk from ancestors to descendants, the genetic links of early life may have been more like a mangrove swamp, with interconnected and tangled roots crisscrossing each other in every direction.

Mayr also examined the processes that give rise to new species—on Mayr's account of *allopatric speciation,* geographical isolation was the most important cause of speciation. For example, a rise in sea level could divide a single large island into many smaller ones. Populations of a single species now isolated on the smaller islands would probably differ from the start just because of the *founder effect:* each would be a small sample of the original population, with a somewhat different distribution of genes. As time passed, more differences would arise, some selected for and some just the product of genetic drift. If the sea level went down again soon enough, the populations would merge once more into a single interbreeding group. But if their isolation continued long enough, even when sea levels fell the populations would be too different to interbreed. At that point, new species have arisen.

Simpson

Mayr's work linked his taxonomy and field evidence about new species with the theoretical work of Fisher, Haldane, and Wright, and Dobzhansky's work on genetic differences in wild fruit fly populations. More and more of biology was being built around the new synthesis. But paleontology remained apart; among paleontologists, intrinsic evolutionary tendencies and orthogenesis were still widely defended. Henry Clay Osborn was a prominent example of this point of view. Like Cope, Osborn argued that the general trends in the ancestry of horses could not be explained by natural selection. Over time the horse lineage lost toes and grew larger, going from dog-sized *Hyracotherium,* with five toes on its front feet and four on the rear to the modern horse family, with only one toe on each foot. The claws at the ends of their toes slowly turned into hooves. Elephants, too, grew larger and larger over time; their teeth became larger and more complex as well. Osborn claimed that these steady changes in particular directions did not fit with natural selection, which would shift direction opportunistically and irregularly depending on environmental conditions and undirected variation. According to Osborn, the earliest members of these lineages already contained the potential to become modern horses and elephants. If the process which reveals this potential was not a Lamarckian striving, Osborn concluded, it must be some other factor we don't yet understand.

It was Osborn's student, George Gaylord Simpson, who brought the new theory of natural selection to bear on the fossil evidence. Simpson examined the lineages Osborn and others had established and showed they were much more complex than the previous workers had recognized. Instead of a single line leading from *Hyracotherium* to modern *Equus,* the horse lineage branched out in many different directions, some growing smaller, some larger, and developing different kinds of hoof.

Simpson also studied the rates of evolution that had been observed in laboratories, and compared them to the rates observed in the fossil record. If

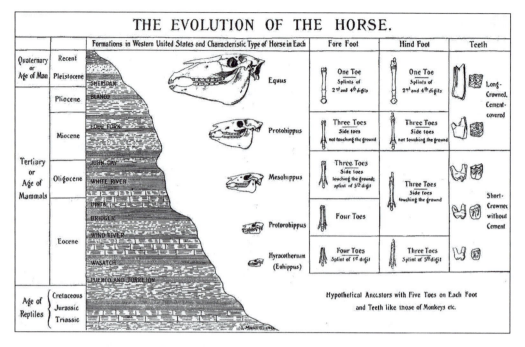

Figure 5.8: A linear view of the evolution of horses. From J. A. S. Watson, *Evolution*. London: T. C. and E. C. Black, 1915. Courtesy University of Calgary Library.

natural selection was the cause of evolution, the rates had to be compatible. To do this work, Simpson had to develop measures of evolution in the fossil record, using detailed features of the bones collected by paleontologists around the world to determine how fast various lineages had changed. The rates of change he found varied both between and within lineages, but they were all much slower than the rates observed in Morgan's fruit flies. Simpson concluded that selection was powerful enough to cause the evolutionary changes found in the fossil record.

Why did it take so long for paleontologists to accept natural selection? Paleontologists like Osborn and his mentor, Cope, made great efforts to trace the lineages of various animals. The risk of this approach is that when we look back in time to find the origins of some present species, it's easy to wind up thinking that the process that gave rise to it was *directed* towards it. As we trace earlier and earlier ancestors, we follow back along a single evolutionary branch. If we turn that branch around and consider how it looks as we move along it from the past back to the present, we may ignore side branches that split off in different directions: after all, they played no role in producing the animal we were tracing the origins of. This linear view of evolution makes it look as though evolution was always aiming to produce the animal whose ancestry we have traced. But this is an illusion. Our focus on a particular end result can distort our view of history. As Simpson argued, what the fossil record really shows is change in a multitude of directions, not systematic progress towards a specific result. Sometimes extinction eliminates all but one branch of descendents, leaving a single form as the only survivor. But this does not mean that this sole surviving form was the goal the entire process was aimed at.

Evolution depends on many contingencies. The horse lineage begins with forest animals. But as grasslands spread, some forms of horse began to specialize at eating grass and living in the open. Running to escape predators rather than hiding in the forests apparently favored stronger and fewer toes, and stronger hoofs. However, some descendants of *Hyracotherium* and its cousins remained in the forests, while still others grew to immense sizes. These other lineages are now all extinct; only the genus *Equus*, including horses, zebras, and donkeys, remains.

Extinction has eliminated all but a single group of horses. In fact, horses became completely extinct where they first evolved, in North America. The ancestors of today's horses survived only in the so-called old world of Africa and Eurasia, where they had migrated. But this does not mean that *Hyracotherium* or some near-

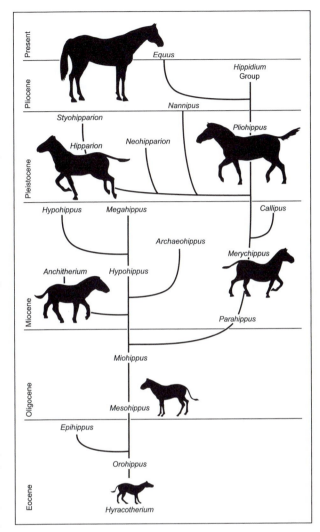

Figure 5.9: G. G. Simpson's more complex branching view of horse evolution. Illustration by Jeff Dixon.

relative of *Hyracotherium* was pre-programmed to produce the modern horse rather than a small forest animal. If the climate had favored forests over grasslands, we might find small forest-dwelling, three-toed horses today instead of our large, one-toed horses. Similarly, if climate changes in Africa had taken a different course, modern humans might never have evolved.

6

BIOCHEMISTRY, DNA, AND THE FUTURE OF BIOLOGY

INTRODUCTION

The modern synthetic theory of evolution was in place by the 1940s. However, our understanding of biochemistry has been completely transformed since that time. In 1953 Francis Crick and James Watson proposed a model of the DNA molecule that revealed the chemical basis of inheritance. Before their model was proposed, genes had been identified only indirectly, by the traits they code for and by their partial reflection in chromosome structure. Crick and Watson's work laid the foundation of a detailed chemical understanding of what genes are and how they work. Their discovery was followed by an explosion of new techniques and new knowledge. Technology and science have advanced in parallel, with advances in each contributing to new ideas and advances in the other.

This new knowledge raises several questions about the modern synthesis, which emerged before the biochemical revolution: how has the synthesis fared in the face of all this new knowledge? Is the modern synthetic theory of evolution still our best account of the history of life when we consider the evidence of today's biochemistry? Or has biochemistry undermined the modern synthesis?

ORGANIC CHEMISTRY

Early in the nineteenth century organic chemistry seemed to be separate from physical chemistry. Organic chemicals could be changed by reactions with non-organic chemicals, but no one had managed to *produce* an organic chemical from non-organic ingredients. Some scientists, called *vitalists*, argued that there was something special about life that prevented organic chemicals from being produced artificially. But the line between physical and organic chemistry was crossed by a young German chemist named Friedrich Wohler in 1828. Wohler used ammonium chloride and silver cyanate to produce urea, which until then had only been found in urine.

Organic chemistry developed rapidly through the rest of the nineteenth century. Scientists began to identify and analyze the main chemicals that make up living things—proteins, nucleic acids, lipids, and carbohydrates. Proteins were shown to be made of carbon, nitrogen, oxygen, and hydrogen, with small amounts of sulfur and phosphorus. These combine to make a range of small molecules called amino acids, which link together in long chains to form proteins.

Nucleic acids (so-called because they are concentrated in the nuclei of cells) were first discovered by Friedrich Miescher in 1869. They combine sugars, phosphoric acid, and five chemicals called *bases:* thymine (T) or uracil (U), cytosine (C), guanine (G), and adenosine (A). Paired chains of the sugar molecules are linked together with phosphate groups to make up the basic skeleton of these molecules, while the bases are strung out in matched pairs along the length of the chain. There are two kinds of nucleic acids: *ribonucleic acid* (RNA), and *deoxyribonucleic acid* (DNA). The differences between these are that they use different sugars (ribose for RNA and deoxyribose for DNA), and RNA uses uracil as a base where DNA uses thymine (a very similar molecule). The order of the bases, strung out like rungs of a ladder along the chains of sugar molecules, is the basis of the genetic code, as Crick and James Watson (with help from Rosalind Franklin) showed in 1953.

Carbohydrates, which include sugars, starches, and cellulose, are used both for cellular structures and as energy stores. These molecules are made up of carbon and hydrogen molecules, together with oxygen. Finally, lipids are based on hydrocarbons, which are generally not soluble in water. Lipids are used for storing energy, in cell membranes, and in some signaling pathways. Fats are one type of lipid, technically known as triglycerides.

The molecules of organic chemistry are unlike any known in inorganic chemistry—they are larger, more complex, and act together in subtle and complex reactions. But they are built out of components that can be synthesized in the laboratory. The vitalists were mistaken; the chemicals life is based on are remarkable indeed, but they are still just chemicals. No line separates the matter of life from ordinary matter.

DNA—THE STUFF GENES ARE MADE OF

In the 1950s, geneticists already knew a lot about how heredity works. But a deep question still had to be answered. Genes were well-established, and chromosomes were known to be their carriers, but no one knew how the information carried by a gene was really stored. This question was answered by Crick and Watson and their successors, who identified the structure of DNA and then cracked the genetic code, revealing how specific sequences of DNA bases determine the corresponding proteins.

Crick and Watson discovered that the normal structure of DNA is a pair of sugar and phosphate helixes joined by pairs of nucleotides (bases) like rungs of a spiral ladder: adenine pairs with thymine, while guanine pairs with cytosine. Each nucleotide base pairs exclusively with its partner base. As a result, each single strand of an unzipped DNA molecule can be used to build a new copy of the original molecule.

The process begins when double stranded DNA is split into two separate strands by breaking the weak bonds that link each pair of bases. Enzymes attach replacement nucleotides, following along one side of the DNA as it unzips and then acting in reverse on the other side. When the process is finished, each side of the divided molecule has been turned into a duplicate of the original double-stranded DNA molecule.

Figure 6.1: Francis Crick and James Watson with their model of DNA. A. Barrington Brown / Photo Researchers, Inc.

Finding that the structure of DNA allowed the molecule to reproduce itself was just the first step towards understanding the genetic code. The next step was to unravel how a sequence of bases encodes the sequence of amino acids that makes up a protein.

DNA uses four bases: thymine (T), cytosine (C), adenine (A), and guanine (G). The genetic code translates a sequence of groups of three bases into the amino acids of the corresponding protein. There are 64 groups of three bases: four possibilities for the first base, multiplied by four for the second, and four again for the third. Each triplet of bases corresponds either to a particular amino acid or to a start or stop instruction for the decoding apparatus (most start triplets also code for an amino acid once the process of translation is already underway).

When proteins are produced, the DNA molecule is unzipped just as in reproduction. But instead of forming new DNA strands from the two sides of the molecule, the unzipped DNA is used as a template to build a *memory* RNA (mRNA) molecule. RNA polymerase attaches to a start triplet on the unzipped DNA molecule, and proceeds to form a complementary mRNA sequence: adenine is matched to uracil (RNA's slightly different alternative to thymine), thymine to adenine, guanine to cytosine, and cytosine to guanine. A cap and tail are then added to the mRNA; in eukaryotes like us, the mRNA sequence may be edited, cutting out certain sections (called introns) and splicing the rest back together. Finally the mRNA is ready for transport to the ribosome, where proteins are produced.

When the mRNA reaches the ribosome, the first three bases of mRNA bind to a site on the ribosome. There the triplet attaches to a complementary *transfer RNA* (tRNA) which carries the amino acid required by the genetic code. The mRNA is shifted to the next triplet, where another complementary tRNA attaches to it, binding its amino acid to the amino acid on the first tRNA. This continues until the end codon is reached and the protein's amino acid chain is finished.

The mRNA sequence has been translated from the mRNA molecule into a sequence of amino acids. The chain of amino acids detaches from the ribosome and folds into its final shape, becoming a finished protein.

The RNA code is arbitrary, in the same way that the word *dog* is arbitrary: nothing about this sound considered by itself, makes it the right word to call dogs by. Similarly, transfer RNAs could be linked to different amino acids, producing an entirely different relation between DNA sequences and the proteins produced from them. This leads to an important observation: different codes are perfectly possible—just as

First Base	Second Base			
	U	C	A	G
U	UUU: Phenylalanine UUC: Phenylalanine UUA: Leucine UUG: Leucine, Start	UCU: Serine UCC: Serine UCA: Serine UCG: Serine	UAU: Tyrosine UAC: Tyrosine UAA: Ochre (Stop) UAG: Amber (Stop)	UGU: Cystine UGC: Cystine UGA: Opal (Stop) UGG: Tryptophan
C	CUU: Leucine CUC: Leucine CUA: Leucine CUG: Leucine, Start	CCU: Proline CCC: Proline CCA: Proline CCG: Proline	CAU: Histidine CAC: Histidine CAA: Glutamine CAG: Glutamine	CGU: Arginine CGC: Arginine CGA: Arginine CGG: Arginine
A	AUU: Isoleucine AUC: Isoleucine AUA: Isoleucine AUG: Methionine, Start	ACU: Threonine ACC: Threonine ACA: Threonine ACG: Threonine	AAU: Asperagine AAC: Asperagine AAA: Lysine AAG: Lysine	AGU: Serine AGC: Serine AGA: Arginine AGG: Arginine
G	GUU: Valine GUC: Valine GUA: Valine GUG: Valine, Start	GCU: Alanine GCC: Alanine GCA: Alanine GCG: Alanine	GAU: Aspartic acid GAC: Aspartic acid GAA: Glutamic acid GAG: Glutamic acid	GGU: Glycine GGC: Glycine GGA: Glycine GGG: Glycine

Figure 6.2: The RNA code. Illustration by Jeff Dixon.

different languages have different words for dogs. But all life shares the same code (though there are some very minor differences). Evolution offers a simple explanation of this fact: any mutation that significantly altered the DNA code would change all the proteins produced by a cell. Such drastic mutations would almost certainly be strongly selected against. So once a successful code was established in an early ancestor, all the descendants of that organism would be stuck with it.

This arbitrary code raises another interesting question: is this code a product of pure chance, a code that life happened to adopt when it first began to manufacture proteins in this way? Francis Crick once took this view, describing the code as a "frozen accident." Recent work raises doubts about this view, though. The code in use is much less error-prone than the vast majority of possible codes. It now looks as though, at a time so early in life's history that it predates the last common ancestor of all life today, the code itself evolved.

Not all the DNA in our genomes codes for proteins. Long stretches of so-called dark DNA appear in the genomes of almost every organism. Much of this is junk DNA, made up of broken genes, broken copies of genes, remnants of genes inserted by viruses, and other useless sequences. These parts of the genome get a free ride when the DNA is reproduced. Because it doesn't do anything that helps an organism to survive and reproduce, junk DNA isn't subject to natural selection. Mutations in these parts of the genome spread or disappear simply by chance, accumulating slowly at a rate that reflects the chance, in each generation, of mutations occurring.

Although sometimes only the DNA sequences that actually code for proteins are called *genes*, not all the important inheritable information is contained in those parts of the genome. Some dark DNA provides attachment sites for switches governing nearby genes. Other dark sequences have been preserved over long periods, protected from the random accumulation of small errors that steadily alters DNA not subject to selection pressures. We have much to learn about how these parts of the genome function, but an important part of the story is their influence on when and where different protein producing genes are active.

THE BIOCHEMICAL REVOLUTION

Understanding how DNA codes for proteins was the beginning of a biochemical revolution. At first, chemical methods for analyzing DNA, RNA, proteins, and other complex organic molecules were very limited. Fred Sanger was the first scientist to sequence a protein molecule. He received the Nobel Prize in chemistry in 1958 for identifying the sequence of amino acids in the insulin protein. The DNA code linking triplets of bases to individual amino acids was worked out during the 1960s; several researchers shared the 1968 Nobel Prize for its discovery.

Today, these first baby steps have led to amazing biotechnical feats. The complete genomes of many organisms have been sequenced, along with particular genes and proteins from many more. We can insert new genes into living things. We can add promoters that modify whether and how various genes are expressed. The key to these technical feats lies in the development of standard techniques and equipment for manipulating DNA. In the 1960s, determining the sequence of a DNA molecule required gently removing bases one at a time and identifying each one. The process required large samples of identical DNA, and it was very slow. Most importantly, it couldn't reliably determine a sequence longer than a few bases. Sequencing the entire genome of a living thing seemed to be an impossible dream—but as time went on, new ways of manipulating DNA have made the dream a reality.

Several fundamental discoveries laid the groundwork for the revolution. One was the enzyme *polymerase*, first found in bacteria. Polymerase adds new bases to single-stranded DNA, turning it into normal, double-stranded DNA.

The process requires a stretch of double-stranded DNA that the polymerase attaches to as a starting point, and it operates in only one direction (from what is called the *3'* to the *5'* end of the strand). Using two primers, short sequences of single-strand DNA that attach to the DNA at a specific site, scientists can select a specific stretch of DNA for polymerase to rebuild. One of the primers attaches at the *3'* (starting) end of each of the separate strands. Polymerase then fills out the rest of the two strands.

Restriction enzymes were another important discovery. These enzymes cut DNA molecules at specific points in the DNA sequence. The cuts are ragged, leaving one end of single-stranded DNA sticking out, ready to bond with a complementary sequence. *DNA ligase* is another key tool; it allows separate DNA strands to be spliced together. When mixed with DNA cut by a restriction enzyme and a new DNA sequence cut in the same way, DNA ligase joins the ends together, inserting the new DNA sequence. Together, these enzymes play an essential role in genetic engineering, cutting and splicing DNA to produce modified DNA sequences.

Gel electrophoresis uses an electrical field to separate different sized strands of DNA. Samples of single-strand DNA are placed in small wells at one end of the gel. Electrodes are attached to the container, creating a field that gradually moves the strands of DNA through the gel towards the other end. Longer strands move more slowly through the gel; after a few hours, strands of different lengths are separated into thin bands in the gel. The DNA strands can be marked using radioactive isotopes or dyes to reveal the bands. This technique has many uses. For example, cutting a particular DNA sequence up with a restriction enzyme will produce a characteristic set of fragments, each with a particular length depending on where the enzyme's target sequence appears in the sample DNA. Applying gel electrophoresis to the results produces a particular pattern characteristic of that sequence—this is a quick way to tell if two DNA sequences are different without having to do a complete sequencing.

Genetic switches were first recognized in the early 1960s, by François Jacob, Jaques Monod, and André Lwoff. The three shared the 1965 Nobel prize for

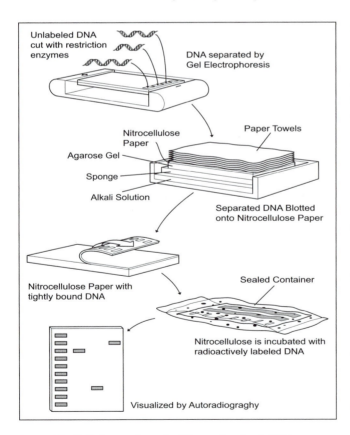

Figure 6.3: Gel electrophoresis. Illustration by Jeff Dixon.

their work on *Escherischia coli* bacteria. *E coli* can produce an enzyme that digests lactose (milk sugars). However, the enzyme isn't produced unless lactose is present. How do the bacteria manage to produce this enzyme when, and only when, the sugar it digests is there? The answer is extremely simple and elegant. The gene is normally blocked by a protein that attaches to a switch on the bacteria's DNA. The protein prevents the gene from being transcribed into mRNA. But when lactose is present, it attaches to the protein and removes it from the DNA. mRNA is then transcribed from the gene in the normal way and the enzyme is produced. So lactose itself activates the genetic switch that turns on the production of the enzyme that digests lactose. There is a large family of proteins that bind to DNA, switching the expression of different genes on and off. Since the discovery of this first genetic switch, these proteins have been found to play a key role in development. They turn genes on and off in different tissues as an organism grows and matures; mutations that alter or duplicate switches can allow a gene to function (or be turned off) in new situations, shifting the patterns of development.

The main method now used to sequence DNA was developed by Fred Sanger, who had already won a Nobel Prize for his work in protein sequencing. Sanger's chain-terminator sequencing technique uses DNA broken up into pieces, each ending in a particular base. The process begins by separating this DNA into single strands. This is done by gently heating it to about 95° C, breaking the weak hydrogen bonds that join pairs of DNA bases. The separated strands are primed to select a starting point for rebuilding the DNA sequence. Polymerase is added, together with bases prepared to fill in the primed single strand. The polymerase attaches the bases to the single strands of DNA, extending them beyond the priming site. But a small fraction of one of the bases is altered so that, once these bases attach to the DNA, the polymerase reaction stops: these bases are the chain-terminators. Using gel electrophoresis to separate the resulting fragments by size reveals all the places in the DNA sequence where the terminating base occurs. Separately using chain-terminators for each of the four bases allows us to determine the entire DNA sequence.

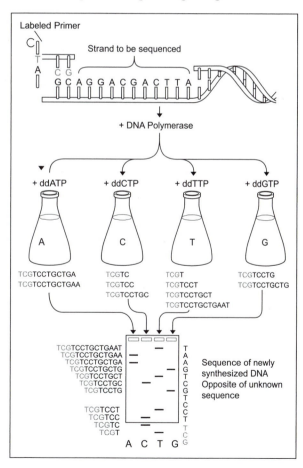

Figure 6.4: Chain-terminator sequencing. Illustration by Jeff Dixon.

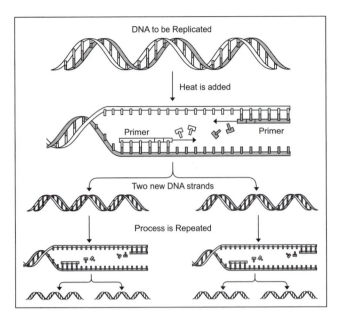

DNA to be Replicated

Heat is added

Primer   Primer

Two new DNA strands

Process is Repeated

Figure 6.5: PCR amplification of DNA. Illustration by Jeff Dixon.

The *polymerase chain reaction* (PCR) is a clever technique that uses the polymerase enzymes that build double-stranded DNA out of single strands to make millions of copies of sample DNA molecules. PCR is especially important because sequencing depends on having large samples of identical DNA molecules. Some kinds of DNA can be cloned and amplified in living systems. But the discovery of the PCR in 1983, by Kary Mullis, made it possible to efficiently copy any kind of DNA.

Polymerase adds bases to single-stranded DNA, converting it into a double strand. Like Sanger's sequencing technique, the PCR uses a supply of bases, ready to attach to a single stranded DNA, combined with *primers*, short strings of DNA that select the starting points for duplication. No chain terminators are used, so the process reproduces the full length of the target DNA. Repeating three simple steps over and over does the job. In the first, the sample DNA solution is heated to separate its two strands (this is called *denaturing* the DNA). Then, in the *annealing* step, the solution is cooled enough for the primers to attach to starting points on the separated strands. At this point the polymerase begins to attach new base pairs. Finally, in the *elongation* step, the solution is slightly warmed again. The primers and the first few base pairs added to them stay in place, and the polymerase works at a faster pace, finally assembling a complete double strand from the separated single strands. The cycle is then repeated: the strands are separated, annealed, and then elongated all over again to form still more double strands. With each cycle the selected DNA sequence is doubled. Normally 20 to 35 cycles are used. This multiplies the original DNA sample by between about 1 million and 30 billion times.

PCR has its limits. The primers must be carefully selected to make sure they bind to the correct region of the sample. Sequences longer than 2 to 3 thousand base pairs are difficult to copy reliably, because the usual form of polymerase produces roughly one error in every 10 thousand base pairs. More specialized forms of polymerase can reduce the error rate by allowing corrections during copying. But in practice longer sequences are usually broken down into shorter ones, linked by overlapping ends.

There are many forms of PCR, used for different purposes. One important example is reverse transcription PCR. This method can amplify, isolate, or identify an already known sequence from a cell or tissues library of RNA. Reverse transcriptase converts the RNA to cDNA. By amplifying the resulting DNA, PCR allows us to determine which genes are being expressed where in an organism. This is one of the laboratory tools transforming our understanding of development today.

Another important application of PCR is in DNA sequencing. Using PCR to produce large amounts of DNA matching a given sample has allowed DNA sequencing on a massive scale. Section by section, linking the sections by overlapping sequences, we have recorded the sequence of base pairs making up the genomes of a rapidly growing number of plants, animals, bacteria, and other living things. Combining PCR with sequencing, using dyes to label DNA fragments and automated equipment to read the resulting sequences allowed scientists to determine the entire human DNA code by 2000.

Today DNA can be divided at selected sites. New genes and control sequences can be added, producing genetically modified organisms. We can alter how genes are expressed by adding promoters, special DNA sequences that make a particular gene more likely to be expressed. We can prevent particular genes from operating by adding small strings of interference RNA that attach to the gene's messenger RNA and prevent it from being processed into proteins.

These techniques can be used to manufacture human proteins in bacterial cultures, including human insulin for diabetics, human growth hormone to treat pituitary dwarfism, granulocyte colony stimulating factor to help cancer patients cope with chemotherapy, and many more. Special proteins that stimulate the production of different types of blood cell are now widely used to help cancer patients cope with chemotherapy. But our focus here is on how this new knowledge fits with our understanding of evolution. As we've seen, there are several ingredients in evolution by natural selection: sources of new variation, the effects of these variations on the organisms that carry them, and the spread or elimination of variations in a population over time. The claim of common descent is another assertion that can be tested against this new evidence.

UNDERSTANDING VARIATION

Many kinds of genetic changes can occur—genes can be altered at a single point or completely duplicated; sections of them that repeat a pattern can be misread, increasing or decreasing the length of the repeats. Genes can be promoted, becoming more active when nearby DNA sequences are altered—or they can be shut down. Whole sections of chromosomes can misalign or be reversed. A one-place deletion in the DNA sequence can completely destroy the functioning of a protein. By shifting the frame of base triplets that are translated into amino acids, these mutations alter the coding for every amino acid in the protein that follows the error, also shifting where the so-called stop codon appears. These things can all go wrong despite proof-reading (in eukaryotes) and other checks on malfunctions, including death of many altered cells.

Some of the mutations geneticists like T. H. Morgan studied have now been identified with specific changes in DNA. One dramatic group of fruit fly mutations causes body parts to appear in the wrong place. Antennapedia, for example, replaces a fly's antennae with legs, while bithorax turns the body segment just behind the fly's wings into a copy of the wing-bearing segment, giving the mutant fly two pairs of wings. Genetic studies showed that the genes causing these strange effects were located on the fly's third chromosome. Five genes on that chromosome were located close together in what geneticists call the antennapedia complex; mutations in these genes alter parts in the front half of the fly's body. Three others affecting parts in the fly's back half were found further along the chromosome. Sequencing of the genes revealed that they all shared a region coding for a sequence of 60 amino acids in the corresponding proteins. This region closely resembles the DNA binding region of the genetic switch protein that controls *E coli*'s lactose-digesting enzyme.

The conclusion was clear: these genes are high-level switches, selecting the developmental paths that different segments of the body will follow. Normally these genes ensure that each segment grows the right parts, placing mouth parts, antennae, wings, and legs where they belong. But when the switches break down, these parts can appear on the wrong segments.

Many mutations that cause human diseases have also been identified. For example, Huntington's disease typically strikes in middle age, producing movement abnormalities, mental and emotional deterioration, and finally death. The normal form of the Huntington's gene produces a protein ending with a long string of the amino acid glutamate; in its mutated form, this string is lengthened. The mutated protein accumulates in brain cells, where it suppresses an important factor that promotes nerve cell growth and affects the mitochondria (the so-called power plants of the cell). The details of how this protein finally kills brain cells are not yet fully understood, but much has been learned, and new ideas for treatments are emerging.

SELECTION

So far we've only mentioned mutations that cause substantial harm to organisms unfortunate enough to carry them. But many mutations don't cause any harm at all. The most obvious examples are mutations that don't change the protein produced by a gene. These synonymous changes replace a DNA triple that codes for an amino acid with a new triple that codes for the very same amino acid. Often the duplication of a gene does no harm either. Duplication can also free the copied gene to change, since damage to one copy still leaves the organism with a working copy.

More interesting for evolution are mutations that actually help organisms survive and reproduce. Mutations that help bacteria resist the effects of antibiotics are an important example. Such mutations include changes in enzymes that break down antibiotics, changes in cell membrane proteins that prevent antibiotics from entering and modifications in membrane pumps that allow the bacteria to eliminate antibiotics before they can do damage.

Tracing the origins of these mutations is complicated by the fact that bacteria can share genes with each other. Some bacterial genes are found on *plasmids,* rings of DNA that bacteria sometimes exchange with other bacteria. Some viruses that infect bacteria can also carry genes from one bacterium to another. So not every resistance gene we find in a strain of bacteria arose by an independent mutation in some ancestor. Still, we have strong evidence that some resistance genes are the result of recent mutations. Penicillin is a natural antibiotic; it was discovered because Alexander Fleming noticed that the growth of bacteria was inhibited near a blue-green mould that had appeared in some bacterial plates. For natural antibiotics like penicillin, resistance can be acquired by exchanging genes with bacteria that are already resistant. But synthetic antibiotics pose a new challenge to bacteria, and resistance enzymes that break them down are the result of new mutations—mutations brought to our attention by the fact that bacteria carrying them survive when their less-fortunate cousins are wiped out. In fact, the stress of being exposed to an antibiotic can cause the mutation rate in some bacteria to increase. Scientists working to keep antibiotic resistance from arising are now seeking drugs that will block the stress proteins that trigger this increase in mutations.

Viruses also evolve, some very rapidly. HIV, the virus that causes AIDS, is a *retrovirus,* a virus that stores its genetic information in less-stable RNA instead of DNA. One of the properties that make HIV such a devastating killer is that it can evolve faster than the immune system can adapt. New strains of HIV with altered protein envelopes evolve before the immune system's antibodies have a chance to eliminate the original form of the virus. As it evolves, HIV continues to attack essential components of the immune system, slowly wearing it down and destroying its ability to fight. These mutations in HIV help protect it from the immune system, keeping the virus alive and enhancing its chances of being passed on to other victims.

Inheritable, selected variation also explains an important part of how our immune system works: how antibodies are formed. When an infection begins, the immune system responds by producing antibodies that attach to the invaders. At first the antibodies are only a very rough match, and the infection often gains the upper hand. But soon more and better antibodies are produced and, at least in most cases, the infection is eliminated. If the same bacteria or virus tries to attack again, the immune system recognizes and eliminates it in short order, using the improved antibodies. This immune system memory is what makes vaccinations work.

How does the immune system respond so rapidly? What enables it to produce antibodies to roughly fit almost any infection, when there are only about 25 thousand proteins coded in the entire human genome? And how does it improve on the initial antibodies, refining them into the weapons it needs to defeat the invaders?

The answer to the first question is straightforward: each antibody producing cell expresses a different antibody on its surface. When it encounters an antigen, a molecule that it binds to, that cell begins to multiply rapidly. This *clonal selection* process quickly produces large numbers of cells that release antibodies against the invaders.

The second question has a subtler answer. The genes used to produce antibodies are actually modified in these specialized immune system cells. Antibodies are *Y*-shaped molecules, including two copies each of two protein chains, a heavy chain and a light chain. The genes for *variable regions* on these chains are built by assembling certain elements from the basic genome of the organism. Three elements are chosen to make up a gene for the variable part of the heavy chain of the antibody; two more are chosen to make up the variable part of a gene for the light chain. There are about 50 choices for the first variable part on the heavy chain. These combine with 23 for the second and 6 for the third to complete that chain. There are 56 more choices for the first variable part of the light chain, and 9 for the second. Altogether, the different combinations of these alternative parts turn inherited codes for about 144 different amino acid sequences into about 3.5 million different basic antibodies.

However, the answer to the third question is the most important point here. How does the rough fit of one or a few of our basic antibodies to the new antigen improve so quickly? The answer is, by random mutation and selection. As new antibody producing *B*-cells develop, the rate of mutation for variable regions of the antibody gene goes up a thousand-fold (some other genes also undergo higher mutation rates). The mutations seem to be random: aside from clustering near the parts of the gene that code for the variable part of the antibodies, they show no special pattern. The cells whose mutant antibodies attach better to the invader's antigen are selected for, and many more cells carrying the successful mutant antibodies (and new mutations of them) are produced. When the infection is past, memory cells retain the most successful of the resulting antibody genes, in case the same infection shows up again.

This is evolution by natural selection, taking place within an individual body. The mutant antibody genes are not passed on to the organism's offspring, since the mutations occur in *somatic* (body) cells, cells that don't contribute to the formation of germ-line cells, that is, eggs or sperm. But this process demonstrates the effectiveness of natural selection all the same. Cycles of undirected variation followed by selection quickly turn an antibody that binds weakly to an antigen into a guided missile that reliably tracks down and eliminates the invader.

The genetics and evolution of more complex features are also being studied. These cases do pose a difficult challenge. Evolution does not respect fixed functions, as Darwin pointed out long ago. Evolution tinkers with parts already present, altering and adapting them to new purposes. It's easy to see how newly evolved antibodies help the body fight infection, but it's much harder to identify how variation and selection worked to construct dramatic adaptations like the long neck of the giraffe, the elephant's trunk, or the large human brain. There are many ways in which a longer neck might have been advantageous for early giraffes. The usual suspect, being able to eat leaves from higher up, is just one possibility; perhaps it had as much or more to do with how the ancestors of giraffes competed for mates. There are also many ways in which a larger brain could have helped our ancestors survive and reproduce, and many ways in which a developing trunk could have helped elephant ancestors in their lives. The uses a part is put to today are often completely different from the uses that drove its early evolution.

This limitation leaves our evolutionary story partly unfinished, but it is not evidence against evolution. In any kind of history, details are left out. Even when we have multiple documents and lots of other evidence about an event in human history, there are still many things we don't know about it. Details of the assassination of John F. Kennedy are still controversial today, even though we actually have film of the event, as well as a detailed record of the investigations that followed and the evidence they discovered. Although we can't expect to reconstruct all the details of what variations arose and how selection acted on them, the evidence for common descent and for natural selection's capacity to construct complex adaptations over time is clear. Some of the details are missing, but we still know that evolution by natural selection can explain the kinds of adaptations we see around us. The familiar stories about giraffes reaching higher and higher into the trees during a famine aren't meant to show exactly how the evolution of the giraffe's neck actually happened. Instead, they illustrate how a simple natural selection process *could have* produced giraffes' long necks.

In fact, long necks affect so much of how giraffes live and what they do that we can't really speak seriously of a single function they serve for giraffes. Giraffes use their necks for many purposes today. Some of these may have been important from the beginnings of their evolution, while some may have become important only recently. The same goes equally for elephants' trunks and humans' brains. What we do know is that variation is always present, that in certain circumstances, selection can consistently favor certain traits (longer necks and trunks, or bigger brains), and that over time natural populations respond to selection pressures by changing their traits.

Darwin is also the source of another important point here: the present variation between different groups and species often allows us to reconstruct plausible paths for the evolution of complex features. For example, many kinds of animals today are sensitive to light. The organs they use vary from simple light-sensitive spots (which allow some single-celled organisms to respond to day/night cycles) to directionally sensitive organs (cup-shaped and lined on the inside with opaque material) to enclosed cups, to primitive image-forming eyes, and so on, in small steps, to advanced vertebrate and molluscan eyes, with irises and pupils, adjustable lenses, specialized central regions for more detailed images, and so on. All these light-sensitive organs contribute to the success of the organisms that possess them, even though many don't form images at all, which is the main use we make of our own eyes. Each step along this long and gradual path is small and potentially valuable to the organism that takes it.

In what follows we'll concentrate on two important kinds of direct evidence for evolution that have emerged from the biochemical revolution. The first is evidence of common descent based on biochemical differences and similarities; the second is evidence for the familiar evolutionary pattern of tinkering with already existing parts, which is now emerging in the study of development.

COMMON DESCENT

The most obvious way to compare life at the biochemical level is to compare genes and proteins from different species to see how similar, or how different they are. At first, this kind of work focused on single proteins and their genes. For example, cytochrome proteins, which play a role in respiration, appear in all aerobic organisms. Cytochrome-c, in particular, is a highly *conserved* protein, present in a very wide range of organisms. Like all proteins, the cytochrome-c protein varies slightly from one form of life to another.

The degree of similarity of cytochrome c from different organisms is a crude but direct measure of how much time has passed since the organisms' most recent common ancestor. The ancestor carried a form of the protein, but changes to the protein have occurred independently along the separate lines of descent that produced the living things of today. Since changes in the protein tend to accumulate slowly over time, groups that share a more recent common ancestor should have more similar proteins.

Evolutionary relationships between different groups have been carefully worked out by studying anatomy, development, and the fossil record. A comparison of cytochrome c proteins matches the pattern of similarities and differences these anatomical and fossil relationships lead us to expect: groups that have been separate for a very long time have quite different cytochrome c, while groups believed to have recent common ancestors also have very similar cytochrome c.

A more dramatic example of genetic similarity arises from a *pseudogene* shared by humans and other primates. Pseudogenes are broken genes which once produced a useful protein, but have suffered a disabling mutation. When a mutation alters a gene beyond repair, it continues in the genome as a pseudogene, accumulating new mutations slowly over time. Because pseudogenes no longer serve any purpose for the organism, these mutations generally have no impact on fitness. As a result, they build up much more quickly than mutations in functional regions of the genome, which are often selected against and eliminated. In fact, the accumulation of this kind of neutral change in DNA serves as a rough sort of clock that we can use to measure the time since two groups separated.

Most mammals produce an enzyme that allows them to make their own vitamin C, so they don't need to have vitamin C in their diet. But primates rarely need to manufacture their own vitamin C. So long as they aren't living on bad food during a long sea-voyage or a bitter winter (things ancient primates never did), they get plenty of vitamin C from their diets. Among primates the gene for a key vitamin C-producing enzyme is broken. The gene is still there in the genome, but it is now a pseudogene, having suffered mutations that prevent it from actually producing the enzyme. One particular bit of damage to the gene is identical in both humans and chimpanzees. The fact that a pseudogene shared by chimpanzees and humans contains the very same alteration in both species is powerful evidence that chimps and humans share a recent common ancestor.

Genome sequences for dozens of plants, animals, and micro-organisms are being worked out today. Three insects (the honeybee, malaria mosquito, and that genetic workhorse, the *Drosophila* fruit fly) have had their genomes fully sequenced; genomes for humans and chimpanzees are also fully sequenced, although work on variations and corrections to some details continue. Genomes for rhesus monkeys, cats, dogs, mice, rats, chickens, and sea urchins are either fully sequenced or in progress, along with rice, wheat, corn (maize), and tomatoes, and a range of bacteria, fungi, yeast, and viruses. One project is even engaged in sequencing the genome of our closest extinct relatives, the Neanderthals, using DNA extracted from a fossil bone and amplified using PCR. Sequences for mitochondrial DNA, which is separate from the DNA found in our chromosomes, are also being determined for many different species.

One of the most widely reported facts in this neighborhood is the amazing degree of genetic similarity between human beings and the common chimpanzee, *Pan troglodytes*. Sequences of our two species' genes match, on average, at over 98 percent of their base-pairs. This result (and similar results showing close genetic resemblance between much more distantly related animals) was surprising because many biologists had assumed that evolution was driven chiefly by the development of new genes. What we have learned—and there will be more to say about this below—is that new genes are not the main drivers of evolution; instead, changes in the *regulation* of genes allow the same basic parts (the different proteins our genomes allow us to make) to be combined in very different ways.

The implications of all this for evolution are clear: tracing relations between living things by comparing genes and proteins reproduces the tree structure familiar from taxonomy. That taxonomic tree was arrived at by comparing anatomy, development, and the fossil record. But it fits beautifully with independent data demonstrating the similarities and differences of multiple genes and proteins. Darwin's idea of common descent, already convincingly supported by the evidence available in the mid-nineteenth century, has been powerfully reinforced by the independent evidence of biochemical similarities and differences that link all of life.

DEVELOPMENT

Multi-cellular animals begin as a single egg cell; their complex structure and organs arise from that one cell. This is an astounding feat—somehow as the organism grows, cells take on different roles in different parts of the body, some producing bone, some becoming muscle, nerves, gut, liver, lung, heart, and so on. They do this in a systematically coordinated way, producing complex structures of cells that differentiate, migrate, organize, and even (selectively) die to form all the organs and tissues of the body. For this to happen, cells playing different roles must express different genes—but what tells each cell which genes to express? How is this process controlled, so that each type of cell and each organ develops where it is needed? Detailed answers to these questions are rapidly emerging, now that we have biochemical tools that can track the process in detail.

The ancient doctrine of preformationism explained this mystery by holding that the adult structure was there from the beginning—in its extreme form, this implied a kind of nesting, with every child already contained in miniature within the sperm or egg of a parent, every grandchild contained in sperm or egg within that miniature child, and so on, generation after generation. The rival theory of epigenesis held that the embryo begins as unstructured matter, and gradually takes form during development. According to epigenesis, structure emerges from the interaction of the embryo's matter with its environment. The debate was difficult to settle. On one hand, the information that structures the growing organism certainly has to come from somewhere; preformationism specifies a simple location and form for the information, while epigenesis relies on interaction between the germ cells and the environment, suggesting that the necessary information arises from the interaction without giving a clear account of how the trick is managed. On the other hand, the indefinite nesting of tiny descendants within their ancestors and the reduction of development to a simple process of growth doesn't fit with the obvious and dramatic changes that take place during development from fertilized egg to adult.

Modern ideas about development have changed the terms of this debate, but they have also led to a similar dispute. Like preformationists, genetic determinists think of the zygote's genetic code as specifying the traits of the adult organism. This was the original idea of what a gene was: some kind of code which specifies particular traits displayed by the organism (blue eyes, five digits on hands and feet, antennae and not legs growing on the head of an insect).

However, the DNA sequences that we now call genes depend on their biological context to produce these traits. The cellular apparatus linking genes to traits includes many proteins, regulatory apparatus that affects which genes are expressed, the cell organs and structures passed on within the egg cell, and many other genes that contribute to producing the trait. In isolation, DNA sequences are as meaningless as the marks on this page would be if no one could read English. Genes do determine a sequence of amino acids. But which sequence depends on the apparatus that interprets that code. Further, whether that amino acid sequence is actually produced depends on complex interactions between the cell, signals from various other cells, and the ways in which gene expression can be suppressed or promoted. Finally, the effects of that single protein, on a cell and on the entire organism, depend on context as well.

Even allowing that the DNA code does contain information about potential amino acid sequences, the information is very different from what preformationists once claimed. It doesn't come from a single parent, and it doesn't come in the form of a tiny, adult-shaped seed. Only by holding its context fixed, taking adequate nutrition, other genes, and the apparatus that regulates their expression for granted, can we treat a gene as *the* cause of a trait and describe it as coding for that trait. At the biochemical level genes are just one cause among many. As the advocates of epigenesis insisted, there is no simple map from structure present in the egg, sperm, or zygote to the structure of the adult organism. A complex interactive process including the entire zygote and its environment is required. The zygote does contain crucial structures, in the DNA of the genome, in the various organelles and proteins of the cell and in the initial asymmetries that begin the process of development and differentiation. But these still depend on a richly structured environment in order to produce a new organism.

Another discovery that emphasizes the importance of interactive processes over genetic determinism is this: the expression of the genetic code can be altered without changing the code itself. Some genes' expression is suppressed by methyl groups attached to the DNA. These groups, made up of a carbon atom joined to three hydrogen atoms, interfere with transcription. A gene's expression can be completely blocked when many methyl groups are attached to it. These groups can also be passed on to offspring, because they tend to be copied when cells reproduce. So some inheritable variations don't require any change in the genetic code at all. Some human genes that cause increased fetal growth are usually suppressed by methylation in egg cells, but active in sperm. There is even a credible selection-based reason for this difference—higher fetal growth is stressful for the female whose body supports the fetus, but it imposes no cost on the male, and could give the male's offspring a better chance at survival.

Until the development of our new biochemical toolkit, biologists studying development had to work on a larger scale. By observing development carefully, they were able to trace the patterns of development in different groups of animals and show how similar they were. Arthropods (including insects, crustaceans, and arachnids), chordates (including all the vertebrates), the echinoderms (starfish and sea urchins), and mollusks are grouped together as *triploblasts*. The embryos of these animals all go through an early phase called gastrulation: the embryo's cells migrate to form three layers of tissue, an inner layer called the endoderm, a middle layer called the mesoderm, and an outer layer called the ectoderm. As the egg yolk is used up, all vertebrate embryos approach the *pharyngula* stage, at which they share a very similar structure. Beyond this point the embryos differentiate, generally following the order of the taxonomic tree as von Baer noted in the nineteenth century: first they develop traits that distinguish their class, after which traits marking order, family, genus, and species emerge.

Observations of various deformities together with experimental work on embryos revealed important patterns in development. Scientists learned how to produce extra toes in chicken embryos and extra eye-patterns on butterfly wings by transplanting small snippets of tissue from specific spots in the developing feet and wings. Clearly, these bits of tissue were producing a signal that organized the development of these parts. But the nature of these signals did not become clear until the biochemical revolution.

By studying which genes are actually expressed in different cells and tissues, scientists have now identified many of the key proteins controlling development and differentiation. Tracing their patterns of expression has revealed a coordinate system telling cells where they are in the developing body. Cells respond to those signals, building all the different tissues and organs in all the right places. One clever way to detect the signals is to light up the key proteins, or the mRNA that produces them. This approach is particularly revealing when applied to organisms with transparent bodies, such as zebra fish. Master genes controlling development at different levels can now be observed at work, laying out coordinates that guide the development of the organism.

THE HOMEOBOX

We've already described the strange mutations antennapedia and bithorax; the normal forms of these genes contain a shared region coding for a protein domain that attaches to DNA and blocks its expression. This shared sequence of codons is called the *homeobox;* the corresponding part of the proteins they produce is the *homeodomain.* The homeodomain binds to DNA, enabling each of these proteins to act as a switch controlling the expression of certain genes. The homeobox genes almost always appear in the same order on the genome, from genes that control development in the front body segments, to those that control the last segments. In insects the homeobox genes are grouped in a single cluster, while vertebrates have four clusters of these important genes, still preserving their head-to-tail order. The homeodomains of mammal proteins are amazingly similar to those found in insects, matching the insect proteins at 58 or 59 out of 60 amino acids.

When part of a protein is still this similar after 500 million years of separate evolution, it must play a critical role in the organisms that use it. All genes that control development on the large scale tend to be very highly conserved. After all, any significant changes in them will change development drastically, and drastic changes in development are very unlikely to produce successful organisms.

Master genes for other features, including eyes, limbs and hearts, are also shared between insects and vertebrates. When the mouse gene *Pax* is inserted and expressed in various places in a fly, eye tissues develop in those places—not mouse eye tissues, but fly-eye tissues. This master gene does not tell the body what sort of eye to build; instead, it simply initiates eye development, both in vertebrates and in insects. Similarly, the *Dll* gene plays a role in building almost any part that sticks out from an animal's body, including limbs in insects and vertebrates and the tube feet of sea urchins. All these proteins include a homeodomain, and they are all more similar to the corresponding genes in other organisms—even very distantly related ones—than to other proteins that include homeodomains.

Recent discoveries about development reinforce several of the key ideas about evolution that we've already encountered. First, new developmental genes and processes don't emerge suddenly and without antecedents; they arise as modifications of already existing genes and processes. We've seen that the same genes play closely related roles in very different organisms. In some vertebrates the usual linear arrangement of the homeobox genes differs, but all these critical controlling genes are conserved parts of the genome. Gene duplication (as we've already seen) is a common kind of variation—duplication events in the early lineage leading to today's vertebrates led all vertebrates to share this basic developmental toolkit. Gene duplication provides new opportunities for selection, as we've already seen. In the vertebrates, more subtle and complex regulation of how cells locate themselves in the body became possible because these new copies of homeobox genes were free to change, adding new layers of developmental control. This kind of variation, building on already established developmental patterns, is just what we should expect from evolution.

Second, new structures don't arise out of nothing either. Instead, they too arise as modifications of already existing parts. The mouth parts, antennae, and legs of insects are all homologous; when the signals that differentiate the specialized parts fail, they turn back into legs. Parts related in this way are called *serial homologues*. Like homologies between different species, they share a detailed common structure, but often they have become specialized, modified in various ways for different functions. One of the most general evolutionary trends is that multiple parts (think here of a millipede, with its many pairs of legs, each pair almost identical to the next) often become both fewer in number and more specialized (think here of the mouth parts, antennae, and legs of a fly). The development of differentiated teeth like our own from simple, repeated teeth (all of identical shape) in early reptiles is another example—as are the vertebrae and digits (fingers and toes) of vertebrates and tetrapods.

The fact that chicken tissues retain the ability to develop teeth is another striking bit of evidence for common descent. Why should the apparatus for tooth construction be present in a bird, when no modern birds have teeth at all? If birds are descended from ancestors who had teeth, it's not surprising that they are still capable of producing them when given the right stimulus.

Third, the master genes that play critical roles in development are highly conserved. Recognizable versions of them appear in vastly different organisms, and the degree of difference between them follows the same tree pattern first noticed in taxonomy. One we haven't mentioned yet is whimsically called *hedgehog*. This gene is named for a mutation in fruit flies, which replaces the usual rows of fine hairs striping the body with an overall bristling pattern like the spiky fur of a hedgehog. There are three versions of the hedgehog gene in vertebrates, one of which is whimsically called *sonic hedgehog*. This gene turns out to be involved in several developmental processes, including the growth of fingers and toes. When it's expressed in the wrong place in a developing hand or foot, it can trigger the development of extra fingers and toes. When its expression or its influence is blocked, much worse things can happen, including *cyclopia*—a drastic malformation of the skull and brain that leaves its non-viable victims with a single, central eye.

The evidence for evolution that we've considered so far in this chapter is more about tracing common ancestry than natural selection. But biochemistry provides strong evidence for selection too. The statistics of gene distributions provide one important kind of evidence here. For example, if two genes are located in separate parts of the genome, then different alleles (variants) of these genes should combine randomly. For example, a certain butterfly gene for wing shape has one allele coding for wings with a long tail and another that produces tail-less butterflies, while a distinct gene for color has a brightly colored version and a dull-colored one. Since these genes combine randomly (as laboratory breeding experiments show), we would expect to see all four combinations, in proportion to the numbers of butterflies carrying each allele. But in the wild we find only two forms: either bright colors and tails, or dull and tail-less. The other two forms are eliminated by selection. They are neither cryptic like the dull, tail-less form, nor do they mimic an unpalatable species like the bright, tailed form. This selection process keeps all four genes in the population, at the cost of sacrificing roughly half of each generation.

When a new variation is strongly selected for, it spreads very rapidly through a population, much more rapidly than a gene that is not being selected for. This is what biologists call a *selective sweep*. We can tell when this kind of rapid spread has happened, because the new gene becomes very common in the population before it has a chance to develop many of the small variations that accumulate from generation to generation. A population that has undergone a selective sweep at some gene locus takes a long time to re-establish normal levels of variation at that locus.

We can also identify the effects of selection in a negative way: regions of the genome that are highly conserved throughout a population, or between two related populations, must have been subject to selection. Variations on these regions do arise, as they do for all genes over time. So if little or no variation remains today, it can only be because the variations have been systematically selected against.

BIOCHEMICAL COMPLEXITY

Some complex biological systems depend on the interaction of several parts to work. Ever since Darwin, critics have argued that this is enough to show that evolution by natural selection is impossible. The idea behind these arguments is quite simple: if a system cannot work without the many parts it now has, there is, the critics claim, no gradual evolutionary path that will allow natural selection to build it. This is because the system cannot work until all the parts are present, but unless it works the parts themselves cannot be selected for, and undirected variation cannot be expected to produce all the parts in one step. Recently, Michael Behe and other advocates of *intelligent design* have applied this old argument to biochemical examples.

Unfortunately for the advocates of intelligent design, the argument just doesn't work. It makes two key assumptions that don't stand up to examination. First, as we've already pointed out, nature does not respect fixed functions. The work that a complex system does now can be very different from the work its parts did when they first arose and were selected for. Second, even if a system now needs all the parts it has to function, this may not have always been the case. The system might have once had parts that allowed it to function in a more direct way, parts that were later lost as more complex pathways took over their role.

One example of this is the eye. Recent work on mice that had lost the light-detecting cells in their retinas showed that working vision could be restored by adding a *bacterial* protein to the nerve cells that normally just transmit the image from retinal cells to the brain. This bacterial protein, rhodopsin, belongs to the family of light-detecting pigments that our retinal cells use to detect light, but it operates in a different, more direct way. Rather than pass its signal along to further molecules, the bacterial rhodopsin directly causes the nerve cells to fire when they're exposed to light. This direct mechanism is cruder than our retinal cells and doesn't allow for as much amplification of the light signal. Still, it does work.

As to how we might have gone from this direct mechanism to the present indirect one, the so-called adaptor molecule involved in the normal visual pathway belongs to a widespread family of signaling molecules involved in many other pathways. Some members of this family already bind to rhodopsin-like proteins. The rest of the pathway also uses proteins that are part of other cell pathways—the whole signaling process could easily evolve by co-opting existing signaling pathways. If the direct signaling pathway were lost after the new, amplified pathway had been established, the final result would require several separate parts to function, even though it evolved by gradual, selected steps.

AND MORE ...

This chapter barely scratches the surface of what we have learned about biochemistry and development. The theory of evolution by natural selection continues to fit the facts, even as what we know about life grows more and more rapidly. Better, it continues to illuminate the facts. Biochemistry, development, and evolution are now intertwined in the field of evolutionary development, or *evo-devo*. There are many places where you can find out more about these fields and their relations, from books, articles, and Web sites for non-experts to research journals where the latest results and discoveries are reported. With powerful new techniques and machinery, more is being learned every day. As our understanding of life grows, evolution continues to occupy the center-ground, linking all life and all of life's processes in a history of descent with modification. The title of Dobzhansky's famous essay rings true today more than ever: nothing in biology makes sense except in the light of evolution.

7

ABIOGENESIS

It is often said that all the conditions for the first production of a living organism are now present, which could ever have been present. But if (and oh! what a big if!) we could conceive in some warm little pond, with all sorts of ammonia and phosphoric salts, light, heat, electricity, &c., present, that a proteine [sic] compound was chemically formed ready to undergo still more complex changes, at the present day such matter would be instantly absorbed, which would not have been the case before living creatures were found.

Charles Darwin (in F. Darwin 1887, 18)

INTRODUCTION

Could life have begun all by itself, as the natural product of conditions on the early earth? Did it? If the answer to these questions is *yes*, biology can be grounded completely in the natural world. Life would have emerged from chemical processes going on somewhere on earth, billions of years ago. The emergence of life from non-living chemical processes is called *abiogenesis*, a word coined by T. H. Huxley. Strictly speaking, abiogenesis is not part of the theory of evolution—evolution is about the history of life, not about its pre-history. Evolution concerns the changes life has gone through over time and what has caused those changes. Abiogenesis, by contrast is a view of how life got started in the first place. But even here, evolutionary ideas play a central role.

In *The Origin,* Darwin made no claims for abiogenesis. Instead, in one of the most widely quoted passages in English literature, he closed his long argument with an appeal to creation:

There is a grandeur in this view of life, with its several powers, having been originally breathed into a few forms, or into one; and that, while this planet has gone on cycling according to the fixed law of gravity, from so simple a beginning endless forms most beautiful and most wonderful have been, and are being, evolved.

(Darwin 1859, 490)

Darwin's hesitation here, like his avoidance of human evolution in *The Origin*, reflects his concern about negative public reaction to a fully naturalistic view of life. However, the close resemblance between human beings and the apes was obvious. His ideas led directly to a clear, plausible, and testable hypothesis about our place in the evolutionary scheme of things. Readers of *The Origin* understood that evolution implied a close relation between us and the apes. Darwin's avoidance of the topic fooled no one.

However, things were different for abiogenesis. Much too little was known for a plausible and testable hypothesis about the origin of life to be advanced. On one hand, it was known that the chemistry of life was not separate from ordinary inorganic chemistry. Many organic molecules had already been synthesized. On the other hand, scientists had come to the firm conclusion that living things do not spontaneously spring into being out of lifeless matter. This now-familiar fact is far from obvious. Given how persistent life is, how small many seeds and spores and eggs are, and how quickly living things seize on available resources, it's not surprising that naturalists had believed for millennia that many living things form spontaneously from non-living matter.

SPONTANEOUS GENERATION

The idea that living things arise naturally and spontaneously from lifeless matter has a long history. Aristotle accepted it, on the basis of what seemed to be perfectly good observations. Shellfish clearly grow out of the mud, with different types appearing under different conditions. Maggots appear in rotting meat, and moths appear in wool that's left around. According to Aristotle, the interaction of moisture and organic matter was the cause of these animals. They could develop where things were rotting, and also inside other animals (which was how Aristotle explained the presence of parasites like intestinal worms). Medieval figures including Athanasius Kircher agreed with Aristotle. In fact, Kircher believed that he had managed to produce frogs from river mud; a recipe for producing small eels from wet cut sods was also reported.

Until more careful observations were made, there was no strong reason to deny that these animals arose spontaneously. One early investigator who put spontaneous generation to the test was Francisco Redi, whose experiments with rotting meat and other smelly things were mentioned in chapter 1. When Redi used light gauze to keep flies off his samples of rotting meat, no maggots grew in it, even though

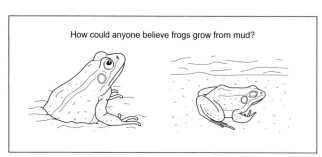

How could anyone believe frogs grow from mud?

Figure 7.1: How could anyone believe frogs grow from mud? Illustration by Jeff Dixon.

uncovered meat was soon full of them. Redi also showed that the maggots eventually became flies of various kinds. Given these observations, Redi concluded that the maggots grew from eggs laid by adult flies, not spontaneously out of the rotting meat. He also pointed out that this relation between flies and maggots had been noticed before, by butchers and hunters who wrapped their meat in cloth to keep flies off, and by Homer in the *Iliad,* in which Achilles worries that flies will lay eggs in the body of his friend Patroclus while Achilles is avenging him. Redi also tried Kircher's recipe for frogs without success—though he admitted that perhaps he hadn't quite got the ingredients right.

Between Redi and other early experimentalists, it soon became clear that, when care was taken to ensure that flies and mice and frogs and eels weren't able to lay eggs or sneak into an experiment, no flies, mice, frogs, or eels would appear. However, a surprising new fact emerged as well: the world is full of much smaller living things, first revealed by Anton van Leuwenhoek's microscope. Perhaps these tiny creatures really did arise spontaneously—but testing this idea would require keeping even invisible organisms out of an experiment. This was a lot harder than keeping out flies and other macroscopic life! In 1765 Lazarro Spallanzani killed the bacteria in a broth by boiling it. He then divided the broth up, putting some into sealed, sterilized jars, and some into jars left open to the air. When bacteria

Figure 7.2: Lazarro Spallanzani tested spontaneous generation by sterilizing and then sealing jars—his techniques led to use of canning to preserve food. From Paul De Kruif, The *Microbe Hunters.* Blue Ribbon Books, 1926. Courtesy University of Calgary Library.

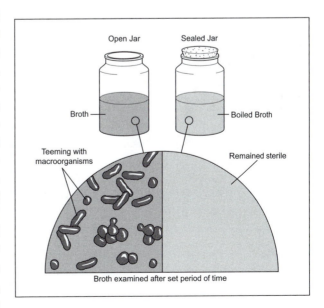

Figure 7.3: Spallanzani's jars. Illustration by Jeff Dixon.

grew only in the exposed broth, Spallanzani concluded that bacteria could not spontaneously generate. But some disputed this, arguing that the air in the jars had been damaged by Spallanzani's process, and that bacteria would develop spontaneously if undamaged air were allowed into the jars. Spallanzani answered some other criticisms with variations on his experiment, but he acknowledged that he couldn't rule out the possibility that the air in his jars had been damaged by heat. Spallazani's process was soon applied to the very practical business of preserving food by sealing it in sterile jars or cans after cooking.

The last defenses of spontaneous generation rested on two questions. The first was about the treatment of air needed to keep it sterile: did treating the air damage the air's ability to contribute to spontaneous generation? Theodor Schwann showed that simply heating air enough to sterilize it before allowing it to come into contact with the broth would prevent the growth of bacteria. But it remained possible that this heating was also enough to damage the air.

The second question concerned an important observation: if hay was steeped in water, the water could be boiled and isolated and still grow bacteria. Félix Pouchet used hay teas for his broth when he showed that even bubbling sterilized air through mercury and into a broth-filled vessel was enough to produce microorganisms. Could a crucial ingredient for spontaneous generation be missing from the other broths, but present in the hay extract?

The first question was answered by Louis Pasteur's famous experiments. Pasteur was seeking to defend his work on fermentation by showing that spontaneous generation played no role in the production of beer and wine. He improved on Spallanzani and Schwann by using jars that were actually left open to the air. Pasteur invented a simple trick to keep dust and spores in the air from reaching the broth: he stretched out the necks of his jars to great lengths, curving and narrowing them. Bacteria failed to grow in Pasteur's

Figure 7.4: Louis Pasteur performed experiments undermining spontaneous generation. From Henry Fairfield Osborn, *Impressions of Great Naturalists*. New York: Charles Scribner's Sons, 1924. Courtesy University of Calgary Library.

jars, and this time no-one could argue that the air in the jars was damaged—perfectly ordinary air slowly seeped into the jars, without being heated at all.

The second question was answered when the Irish physicist John Tyndall showed that some bacterial spores could survive the boiling process. He went on to develop a new process for sterilization using a series of relatively gentle heatings in place of one intense one. Spores that survived the first heating were stimulated to begin growing after the broth cooled. This made them vulnerable to the next cycle of heating. Tyndall's process prevented even hay extracts from growing bacteria under sterile conditions.

But Tyndall's most famous con-

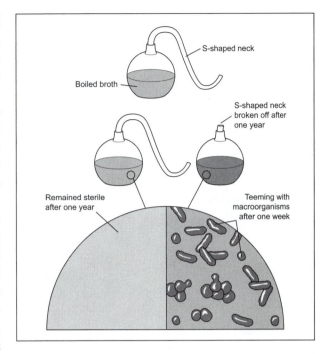

Figure 7.5: Pasteur's jars were open to the air but protected from contamination by their long necks. Illustration by Jeff Dixon.

tribution to the debate over spontaneous generation was an improvement on Pasteur's long-necked jars. In his work on optics Tyndall had used a beam of light to illuminate dust particles in the air, studying how the particles scattered the light. He had also investigated air from deep in his own lungs, suspecting that the many narrow passages in the lungs would capture and remove any particles of dust. The air from his lungs contained no dust particles at all, as he showed by shining a light through it. Tyndall went on to build a box whose interior he coated with glycerin. Air could get in and out of the box through convoluted glass tubes. The box's front side was made of glass for viewing; it also had round glass windows on opposite sides to shine light through. After closing the box up and leaving it for three days Tyndall shone a bright light through the windows to show that no dust particles were left to scatter the light; as he put it, the air inside the box was optically empty. Tyndall went on to use the box in series of experiments, showing that sterile broth, urine, meat, fish, and vegetables did not spoil when kept in his box. The results of his work appeared in 1870; the long debate over spontaneous generation was finally over.

The demise of spontaneous generation added more urgency to efforts to improve public sanitation. Having realized that germs cause disease, and that germs don't arise spontaneously but instead are spread through dust and dirt, in contaminated water and air, and on dirty hands, it seemed clear that disease and infection could be prevented by using antiseptics in the operating room, by hand washing, by ensuring access to clean water, by pasteurizing milk, and by implementing other public hygiene measures. These measures have paid

Figure 7.6: John Tyndall, physicist and clever experimenter. From John Tyndall, *A Library of Universal Literature: Part 1: Fragments of Science.* PF Collier and Son, 1900. Courtesy University of Calgary Library.

huge dividends for human health. For the first time in history, the vast majority of children in the developed world actually lived to adulthood.

ABIOGENESIS

Spontaneous generation in the traditional sense—a common process going on all around us—was scientifically untenable by the late nineteenth century. But the idea that life might arise from non-living matter under the right conditions survived. As Darwin pointed out in his famous letter to Hooker, a broth of organic chemicals exposed to today's world would quickly be consumed by the life already present. But this fact leaves the big questions open: what might happen to Darwin's warm pond if it were left alone for many years in a world that did not yet contain life? Were there other processes and places on the primitive earth where organic chemistry could get started? Could life somehow arise out of a complex set of chemical processes?

The fossil record contains evidence of life from very early in the earth's history. Despite the odds against such minute living things leaving any trace at all and the immense difficulty of finding those traces, microfossils of bacteria have been found in rocks dating back as far as 3.5 billion years ago. Less direct evidence of still earlier life appears in ratios of carbon isotopes from 3.8 billion year old rocks in Greenland. Living things take up more of the lighter isotope carbon 12 than of carbon 13; consequently, carbon from organic remains contains a higher percentage of carbon 12 than inorganic carbon does. The ratio of carbon 12 to carbon 13 in these ancient rocks strongly suggests that it was processed by living things.

Life appears to have emerged just after the earliest part of earth's history, a period so violent and inhospitable to life that we call the *Hadean*. In that long-ago eon, the earth was still in the throws of being formed. It is hard to imagine, from our comfortable perch on this now-friendly planet, just how violent this time was. Large asteroids and comets repeatedly struck the earth, including one late, immense catastrophe: an impact actually broke up the planet, creating the moon. Life would have had a very hard time surviving these massive bombardments. But the earliest evidence of life appears in rocks laid down just after this inhospitable beginning.

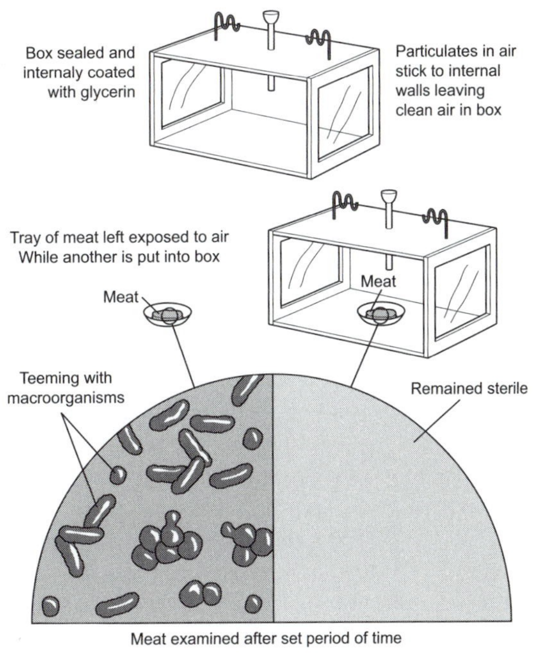

It follows that conditions somewhere on the early earth must have gotten life started almost as soon as they possibly could. But what were these conditions? John Tyndall coined the word *panspermia* to describe the near-universal presence of life on earth, floating on dust particles in the air. This meaning was extended by later

Figure 7.7: A Tyndall box. Note the convoluted tubes that allow air to flow in and out, but catch dust particles. Illustration by Jeff Dixon.

scientists, for whom panspermia became the idea that life floats through the entire universe in the same way. These scientists suggested that life had arrived on earth as colonists, drifting in on cosmic dust. From this point of view, of course, the only conditions necessary for life to get started on earth were that the earth become habitable for some of the life forms floating in from space.

But panspermia only pushes the question of life's origins back a step. If life arrived here from outer space instead of beginning on earth, we still want to know how life began. What conditions, what processes could have taken us from the chemistry of minerals, gases, and water on some lifeless world to the reproducing, evolving entities we call life?

Probability, Randomness, and Silly Calculations

Skeptics about abiogenesis have offered probability arguments against the idea that life could ever arise from non-living matter. Typically these arguments focus on the probability of features of living things today: a single protein, for instance, may be a string of amino acids 200 or more units long. The probability of producing that exact protein by randomly assembling amino acids will be astronomically small: there are 20 amino acids in general use by living things. Assuming that our random assembly procedure makes each amino acid equally likely to occupy any of the 200 slots in this protein, the probability that a randomly generated protein will match our target is 1 in 20 for the first amino acid in the string, 1 in 400 for the first two, 1 in 8,000 for the first three, and 1 in 20^{200} for the entire protein. This is a staggeringly unlikely outcome, so unlikely that there isn't enough time and material in the universe to even come close to performing enough trials of the random procedure to produce the desired outcome.

But of course this isn't how that protein came into existence. It is not the product of random assembly of amino acids. At the level of basic biochemistry, things are fully deterministic: any run of Miller's experiment will produce the same amino acids and other organic molecules. Even when we begin to consider complex situations where the chemistry may be too complex to predict, like Russell and Martin's iron-sulfide bubbles, the basic principles of chemistry still apply as molecules are produced with the help of the mineral catalysts and flows of small input molecules.

As the process develops, some degree of randomness may come in—it may be, for instance, that some process for building RNA out of nucleotides really doesn't care which nucleotide it adds next, so the odds of a particular nucleotide appearing in a particular location on the molecule is just a matter of probability, roughly equal to each nucleotide's proportion in the local population of nucleotides. But no one expects this random process to produce (for instance) a particular version of a modern transfer RNA molecule.

What we do expect to happen in the RNA world is that some self-replicating cycle will get started. In vitro variation and selection has already produced a range of interesting ribozymes (RNA catalysts). This implies that there are a lot of RNA molecules that can do these jobs. So it's not at all surprising that ancient living things found molecules that could carry out the basic functions of life. Better yet, because there are so many molecules that can do these jobs, when we see very similar molecules doing these jobs in very different living things, that's evidence that these living things share a common ancestor that made use of a similar molecule.

If you're still worried about how that modern protein came to be the exact protein it is, don't be. In a way, it's just like a lottery: it's extremely unlikely, for each person who buys a ticket, that that person will win the lottery. Still, given that a lottery occurs, someone will win. So sometimes it can be very probable—even certain—that some highly *improbable* event will occur (but only when we really don't know which improbable event it will be). This may sound paradoxical, but that's only because we confuse the probability that *some member or other* of a very large class of events will occur (which can be high) with the probability of a *particular* member of that class of events (which can still be vanishingly small).

For instance, if I decide to flip a coin a thousand times (tedious, but doable) when I'm done I will have produced a particular sequence of results—HHTTTHTHHHHTHTTH and so on—that has the fantastically low probability 1 in 2^{1000}. The same point holds for proteins. Any particular sequence of 200 amino acids is extremely unlikely. But once life began translating RNA codes into proteins, it was very probable that 200 amino acid proteins would occur. Any such protein has to have some particular sequence of amino acids. So it was highly probable that some extremely improbable result or other would occur. No miracle is required.

You may also come across arguments claiming that complex systems cannot develop naturally because this would violate the second law of thermodynamics. This is just false. The second law of thermodynamics describes what happens to a system in isolation—in such systems natural processes always increase net entropy. But no law says that a system through which matter and energy are flowing must undergo an entropy increase. If it belongs to a larger isolated system (the universe as a whole, for example) the larger system must experience an *overall* entropy increase, but parts of such systems can decrease in entropy—otherwise, familiar phenomena including growth, refrigeration, and internal combustion engines would be impossible.

TOP-DOWN OR BOTTOM-UP?

How should we approach this question? There are two general strategies: the first is to begin by looking at life as it is today. By identifying its most universal and indispensable components we may be able to come up with a minimal recipe for modern life. This approach won't do the whole job for us. Even at its most basic, modern life is extremely sophisticated and complex compared to what scientists believe the first self-reproducing organisms were like. But the top-down approach helps by identifying key chemical processes involved in all modern life. We can continue to work downwards from these by examining these basic processes and considering how they might have evolved over time from still simpler processes.

We can also identify basic materials and energy sources on the early earth and study the chemical reactions that can occur given with these as inputs. If the products of these reactions turn out to be molecules that are important to life, then we've identified one way to produce some of the building blocks of life. Again, this won't do the whole job. We will need to do a lot more work to figure out how these building blocks could come together to produce something we can truly describe as alive—that is, something that draws on energy and other materials in its environment to maintain and reproduce itself.

When you're working on a complex problem, it's always a good idea to work in both directions. Working from the endpoint or goal—whatever you are trying to produce—can clarify your understanding of the goal and some of the processes that could have led to it. Working from possible starting points—the simple elements with which you must begin—can reveal new interactions and unexpected possibilities. Research on abiogenesis is proceeding in both directions today. Biologists are identifying the most basic processes of modern life, and exploring how they might have come from still simpler systems. Others are studying how simple chemistry on the early earth could have built up the materials needed for a first, simple form of life.

In the end, the two approaches must be joined in the middle to give us a path leading from simple chemistry to life as we know it. We will probably never know whether any particular path or paths we find is the right one, the true path that life on earth actually followed. The evidence to settle that question has almost certainly been destroyed in the course of time. Some clue, some wisp of a remnant that we could eventually decipher may remain—it's always risky to rule such possibilities out absolutely, since science can surprise us. But it seems very unlikely, and it's not crucially important, either. The demonstration of a possible path, or perhaps of more than one, will be accomplishment enough: for the first time, life itself will be fully grounded in our understanding of the natural world.

Finding a path that reaches the point of reproduction is clearly the key. As we've already seen, when reproduction is underway, selection begins—any heritable advantage belonging to a new form tends to spread through the population. As new and advantageous variations arise and spread, life changes. Old chemistry is shifted and altered to perform new tasks, new and different environments are occupied, new resources are encountered in the environment and incorporated into the operations of life. Once life is underway, evolution takes over.

OPARIN AND HALDANE

In the 1920s a more detailed idea along the lines of Darwin's warm pond took shape. The oxygen in our present atmosphere tends to break down organic molecules. It's hard to imagine large amounts of organic material building up under these conditions. But the present atmosphere with its high oxygen content is a *product* of life. The geological record shows that oxygen concentrations were much lower in the past—for example, the high levels of oxidized iron in Triassic rock indicate that a dramatic increase in oxygen levels took place at that time.

Before life began, the earth's atmosphere would have been very different. In particular, it would have included very little oxygen. In what scientists call a *reducing* atmosphere, made up of reactive gases like hydrogen, methane, and ammonia, along with carbon dioxide and the relatively inert nitrogen, many processes producing organic chemicals can occur, and the breakdown of those chemicals would be much slower than it is today.

This general idea was proposed by a Russian biochemist named Aleksandr Oparin in 1924. A few years later the brilliant English biologist J. B. S. Haldane made a very similar suggestion. Oparin's idea was that carbon, freed from the rocks of the cooling earth, would combine with hydrogen to produce various hydrocarbons. Nitrogen and hydrogen would react to form ammonia, which in turn would produce cyanide and amino acids. This chemical process, he proposed, would eventually have given rise to a rich variety of interacting chemicals, out of which life itself could emerge.

Haldane emphasized that an ancient atmosphere with low oxygen levels would produce very little ozone, and the lack of ozone would allow much more ultraviolet radiation to reach the earth's surface. This radiation would provide an energy source for a wide range of chemical reactions, helping to produce sugars and amino acids from raw materials including ammonia, water, and carbon dioxide.

Oparin continued to work on these ideas throughout his life, publishing two more books on the subject. He suggested many detailed chemical pathways for producing different organic chemicals. He also proposed an important role for membranes in the early chemical steps leading to life. Fatty molecules in water can group together to form droplets; sheltered inside these droplets, chemicals can interact while being protected from other chemicals in the water. A kind of selection can begin if some chemicals act to stabilize the droplets they are in. Some droplets might even split in two, in a kind of crude reproduction. Oparin suggested that these processes of chemical selection and reproduction were the starting point on the long path to life.

STANLEY MILLER

The first lab experiment designed to test Oparin's idea of the so-called primordial soup was carried out by Stanley Lloyd Miller in 1953. At the time,

Miller was a graduate student at the University of Chicago, working under the direction of the Nobel laureate chemist, Harold Urey. Miller had been inspired by a lecture Urey gave on the early chemistry of the earth's atmosphere. Despite Urey's advice that he should pursue a more modest project with a higher probability of success, Miller wanted to try to produce a laboratory model of Urey's ideas. Urey set a time limit on the project: if results did not emerge within a year, he wanted Miller to give it up and turn to a less speculative and risky project.

The apparatus Miller and Urey designed was extremely simple—a closed loop with two flasks connected by tubing. The larger flask had two electrodes near its top, between which a spark would jump. The tubing ran from the top of the small flask up, across and down into the top of the larger flask. More tubing ran out the bottom of the larger flask, through a u-shaped trap and back into the small flask. The air in the apparatus was carefully pumped out and replaced with a mixture of hydrogen, methane, and ammonia. The small flask and the trap were filled with water. Then the spark was turned on and a low flame applied to the bottom of the small flask.

After two days the water had turned pale yellow, and black tarry deposits had formed around the electrodes. Miller ended the first run and began testing the water for amino acids. Dipping the tip of a paper filter into the solution, he let the solution creep up into the paper. Because different molecules move upwards at different speeds, this test spreads different chemicals in the solution across different parts of the filter. Drying the filter and then treating it with a developer created an image on the paper of the various chemicals that have been drawn up into the paper. The developing chemical was ninhydrin, which reacts with amino acids to produce a range of colors from yellow to purple. Miller was delighted to find a characteristic purple spot on the filter, just where the simplest amino acid glycine should appear.

A longer run, this time with the water boiling, produced more dramatic results. Now

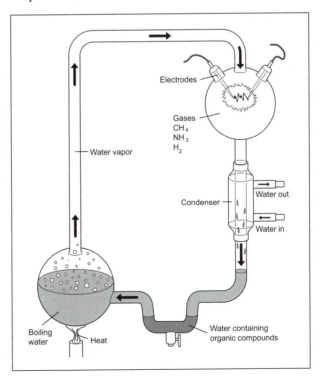

Figure 7.8: Miller/Urey Apparatus. Illustration by Jeff Dixon.

several more amino acids appeared on the developed filter. Further experiments soon followed. In 1961, Juan Oro showed that amino acids could also be produced from ammonia and hydrogen cyanide in a water solution. Oro's experiment also produced large amounts of adenine, one of the nucleobases in RNA and DNA, and an important component of adenosine triphosphate (ATP), the key energy currency of cellular metabolism.

Since his initial experiment Miller and others have studied many variations on it. In one or another of these experiments, 17 of the 20 amino acids that make up proteins have been produced, along with all the nucleobases appearing in RNA and DNA. However, such experiments have not produced the sugars that form the backbone of RNA and DNA; sugars seem to require different conditions.

The composition of the early atmosphere is still controversial. Some scientists believe it was different from the mixture Miller began with, although recent work suggests it may have been very similar to the Miller-Urey mixture. In any case, oxygen was scarce in the early atmosphere, and hydrogen was far more common. Both conditions make the formation of organic molecules easier. A range of such test atmospheres have been tested in the lab, with results similar to Miller's.

The early earth had other potential sources of organic molecules as well. One that has generated intense interest is hydrothermal vents. In 1977 spectacular life forms were found clustered around undersea hot springs along the Galapagos rift. Minerals dissolved in the water (which emerges at 600 degrees Fahrenheit) form tall chimney deposits called black smokers, because of the murky water that spews from their tops. Bacteria feed on high levels of hydrogen sulfide in the waters that rise from the springs. Other animals, including clams, mussels, tube worms, and deep-sea crabs, feed on the bacteria and each other. The entire system is independent of the sun's energy, relying instead on the chemical energy rising from the vents.

Could the first life have developed around these vents, and turned to the sun only later? The debate over this idea has been vigorous. On one hand, the high temperature of vent waters actually tends to break down organic molecules. On the other hand, interesting chemistry does happen as the water emerges and cools in the high pressures of the deep sea. Perhaps some of the chemicals would survive by escaping from the destructive conditions in which they formed. The chemistry of these vents represents another possible source for interesting organic molecules on the early earth. In addition, the vents spew out large quantities of metal and mineral ions. These may have played important roles in chemical reactions going on elsewhere.

At the other extreme, the polar ice caps offer another interesting environment for early organic chemistry. Beneath the ice, molecules would be protected from destructive ultraviolet radiation. They could slowly accumulate, forming the proverbial primordial soup.

Another surprising potential source of organic molecules is outer space. A wide range of organic molecules have been found in carbonaceous meteorites. It turns out that these chemicals can survive the meteorites' fiery descent through our atmosphere. The molecules themselves are very similar to those produced in Miller-type experiments, which shows clearly that the kinds of chemical synthesis Miller managed to achieve in his experiment will occur under a wide range of conditions. In one experiment exploring possible synthesis of organic molecules in space, scientists placed an aluminum disk in a vacuum chamber. They deposited frozen gases—water, ammonia, carbon monoxide, and methane—on it, and then exposed the disk to UV radiation. A wide range of interesting organic chemicals were formed—chemicals matching those detected in large, thin clouds of gas and dust in outer space. So meteorites and comets could have constituted an important source of organic molecules on the early earth.

Finally, recent work on a less-dramatic kind of hot spring offers interesting hints of how complex biochemical systems might have arisen. At high pressures but less extreme temperatures than those of black smokers, some simple minerals including iron and nickel sulfides can catalyze important biochemical reactions—reactions that depend on complex enzymes in modern living things.

WORK IN PROGRESS

Important puzzles remain. One is that, while the kind of synthesis explored by Miller and others can produce many amino acids in a kind of watery soup,

Laboratory Procedures and Good Science

In *Gen-e-sis* (2005), a non-technical book on the origins of life, Robert M. Hazen describes his work with Glen Goodfriend, a chemist who helped him with a challenging experiment. The aim of the experiment was to demonstrate a chiral selection effect: because different faces of a crystal actually have different handedness, Hazen proposed that they might preferentially adsorb either the right or left-handed versions of an amino acid. What is most striking about this story is the meticulous care that Goodfriend took with their laboratory procedures. Every step of the operation was set out in detail; timing, temperature, the movement of samples from one stage of the procedure to the next—all these variables were measured and procedures established to ensure that every sample was treated, as far as was humanly possible, in exactly the same way. Even very small deviations were noted. Taking care of these details was the only way to keep the inevitable variation from one sample to another low enough for the effect they were interested in to be detected. Finally, the samples were randomly re-numbered by Hazen, and again by Goodfriend, to ensure that neither would know which samples were which until the results were all in. Establishing and documenting such detailed, repeatable procedures is an essential part of the craft of science; carefully ensuring that results are not biased by experimenters' expectations is another. Not every question can be addressed in such a systematic, careful way—but scientists are expected to do whatever they can to meet these standards.

these amino acids are difficult to link together. In water, long chains of amino acids tend to split apart. One solution to this problem is to use a surface that the growing molecules can attach to. Minerals and clays provide templates where amino acids can bond together to form peptides (short chains of amino acids) and proteins (longer chains).

Early work on this problem of linking amino acids together was done by Sidney Fox. His experiments on solutions of amino acids involved drying them out on heated surfaces and studying the results. His process generated complex tangles of linked amino acids that Fox called *protenoids*. Though very untidy compared to modern proteins, these molecules had some interesting properties. Some could catalyze reactions, though not as efficiently as protein enzymes do today. Strikingly, in water they formed spheres which could grow by absorbing more protenoids, and sometimes even divided in two. But Fox's later ambitious claims that protenoids were the solution to the problem of life's origin were seen as overblown: protenoids are too messy, with their many branches and crosslinks. To other investigators they looked too random and unstructured to provide a basis for life. Also, Fox's work on protenoids provided no hint of how RNA and DNA became involved in reproduction and the manufacture of proteins. Still, protenoids did illustrate a kind of primitive metabolism, based on the cycles of drying and polymerization that produce protenoids from amino acids. Immersion in water can lead the protenoids to form spheres and expand already existing protenoid spheres. And protenoids are chemically similar to proteins, even though they are structurally much messier.

Other work, especially by A. G. Cairns-Smith, has examined clay as a possible part of early life forms. Clays may contain reproducible patterns in their own right, in the form of crystal patterns and flaws that are repeated as new layers of clay are formed. Some of these patterns may reproduce more successfully than others, spreading through a clay formation, and accidental variations on the more successful may improve on their success, too. As we've already mentioned, clay surfaces can provide templates on which larger organic molecules can form. So such clay-based evolution could even have led to a transfer of information from clay structures to more familiar biochemistry.

However, the experimental challenges here are daunting. Observing the formation of new clay layers and figuring out how their features depend on those of the underlying layers is extremely difficult. Deciphering which features of the layers matter and testing ideas about them would require much more detail and precision. Finally, bridging the gap between clay as DNA and the present role of RNA and DNA is a major challenge. Nevertheless, the idea of an early, crystal-based replicator remains an elegant suggestion for how life began.

MIRROR IMAGES

Another important puzzle about the chemistry of life turns on a peculiar fact about organic molecules. In many cases, the same compound (made of the same elements, connected to each other in the same ways) has two distinct forms which are mirror images (enantiomeres) of each other. Louis Pasteur was the first to realize this, connecting it to the power of pure solutions of each form to polarize light in different directions. Artificial synthesis produces each form in equal amounts. But there is something special about the forms of these molecules found in living things: living things make and use one form almost exclusively. Proteins are made only with the left-handed forms of amino acids, while all sugars produced by living things are right-handed.

Understanding how one form came to be selected when both are so similar is a major challenge. A number of processes, including interactions with polarized light and absorption onto crystal surfaces, can distinguish the two forms. At a more fundamental level, there is an asymmetry between left and right in the process called *beta decay,* in which a neutron emits an electron and turns into a proton. This process emits polarized radiation, which might favor one handedness over another in the chemical reactions it powers.

But perhaps no general process favored the forms life uses—perhaps it was just local conditions that happened to choose one over the other. In present day life enzymes at the very beginning of the process that forms sugars select for handedness. The rest of the process just inherits choice made at the outset. When it comes to life's origins, once a choice was made, even if just by chance, life would have been stuck with it: proteins go together properly when they are built from amino acids that share the same handedness, since that ensures that the corresponding chemical groups on each amino acid are aligned together. A recent experiment suggests that beginning with a slight difference between one form and another can lead to a dramatic difference in their concentrations in a solution. So an accidental choice could have been locked in very early in the history of life.

We have just begun to explore the range of chemical processes that could have been underway on the early earth. A planet is a very large place, and conditions on the early earth varied widely, just as they do today. We don't yet know all the processes that may have been involved; we may never identify them all. Neither do we know just how the gaps between the synthesis of bio-chemical building blocks and functioning organisms were crossed. But there are many possibilities to explore.

REPRODUCTION VS. METABOLISM

Which came first, reproduction or metabolism? Metabolism includes all the processes by which living things use energy and materials available in their environment to grow and develop. Reproduction, of course, includes the processes that allow living things to make copies of themselves, including their metabolic systems. Both are essential to life as we know it. Some basic chemical processes involved in metabolism and reproduction are common to all living things. Presumably they were present in the common ancestor of all today's living things. But did one of these somehow start the process that led to life, only later recruiting the other? Did so-called naked reproducers begin by drawing directly on materials in their environment, and only later learn to build metabolic systems that would provide the materials more reliably? Did metabolism start the ball rolling, producing the basic materials and chemical cycles out of which reproduction finally got started? Or did both somehow manage to develop together?

Self-Reproducing Molecules

One goal of the reproduction first approach has been to identify molecules that can actually make copies of themselves when the right raw materials are available. A few such molecules have been identified. These molecules have to act simultaneously as catalysts and templates for a reaction that produces the copies. In the simplest case a self-reproducing molecule will act on two other molecules, combining them to make a copy of the initial molecule.

A more complex system could involve two molecules, each of which helps to produce copies of the other from the available raw materials. Or there might be a group of molecules which act together to produce more molecules belonging to the group. Any collection of molecules like this will go on reproducing so long as the input molecules and energy required for the reactions are available.

One advantage of the reproduction first approach is that as soon as an imperfect reproducing system arises, natural selection begins. Any variation in the system that improves its ability to persist and reproduce will spread and eventually replace inferior forms. Even slight advantages add up quickly. As soon as reproduction is underway, life will begin to change and improve, expanding and refining its ability to use the resources around it to make more copies of itself.

The RNA World

In the early 1980s, scientists discovered that RNA can do more than just store information in its sequence of bases—it can also catalyze chemical reactions. These active RNA molecules (called *ribozymes*) suggested that RNA might have been a jack-of-all-trades at the beginning of life, carrying the information that gets reproduced while also driving the reactions needed for its own reproduction. Some very exciting work has used selection processes in the test tube to produce RNA molecules that can carry out various chemical tasks, such as catalyzing particular reactions or acting as templates to bind sequences of other molecules together. Some of these RNA molecules are self-reproducing, making copies of themselves from a feedstock of chemical parts. This work is a perfect example of the reproduction first approach. Once an RNA molecule (or a collection of such molecules) began to catalyze reactions that produced more copies of itself, selection would begin. The climb towards life as we know it would be underway.

This proposed stage in the history of life is sometimes called the RNA world. Current pictures of this world come in several versions. Naked RNA replicating itself from the raw materials of bases and ribose sugars is the simplest version. But RNA might have been secluded behind some kind of membrane, open enough to allow raw materials in, but confining enough to keep the RNA and other large molecules inside. RNA might also have been attached to a mineral surface. The surface could stabilize the RNA (or even be the template on which the RNA formed) while still allowing the RNA to interact with chemicals in the surrounding water. Such a surface-bound RNA might have produced peptides (short forms of proteins) directly, because different RNA sequences have affinities for particular sequences of amino acids. Finally, the surface-bound RNA might have acted on nucleotides themselves to produce more copies of itself.

RNA is far less stable than DNA, so it's not surprising that the task of data storage was eventually turned over to DNA. But RNA is still the active molecule in protein manufacture, guiding the process of reading information from DNA and then building the indicated proteins. RNA also plays a role in shutting down some genes and activating others. RNA molecules can even prevent some proteins from being produced by interfering with the RNA that codes for them. RNA plays these roles in all living things today, from eukaryotes like ourselves to bacteria and archaebacteria. The central role RNA plays in so many processes shared between so many living things suggests RNA may well have done these jobs in the common ancestor of all life. This conclusion is especially convincing because in many cases there are other chemicals that could do the job, so there is no reason to suppose that different forms of life would independently come to use RNA in all these ways. The fantastic versatility of RNA is the main motivation for exploring the possibility of an early RNA world. It seems that once upon a time almost all life's processes, from information storage and reproduction to chemical catalysis, might have been carried out by RNA.

An important turning point in the RNA world story is the shift from RNA catalysts (ribozymes) to protein catalysts (enzymes). From the outset, RNA could have produced strings of amino acids. Some of these would have had useful chemical properties, but the ability of RNA to directly produce a particular sequence is quite limited. In modern cells RNA sequences specify a sequence of amino acids indirectly, by using a chemical translation process. But RNA on its own can select amino acids directly. Some sequences of RNA will bond to hydrophobic amino acids while others prefer hydrophilic amino acids. Even this crude control would allow RNA to partly determine the shape of the resulting peptide or protein, because a protein folding in water tends to keep hydrophobic amino acids on the inside and hydrophilic ones on the outside. Some RNA affinities for amino acids are even more particular—for example, the sequence of bases AAA (triple adenine) bonds especially well with lysine. Interestingly, this triplet still codes for lysine today, despite the fact that the modern translation apparatus would allow the triplet to pick out any amino acid at all.

The later development of translation apparatus (now shared, with very slight differences, by all living things) was a critical breakthrough. Translation ensures that each sequence of RNA (once prepared for protein construction) corresponds to a specific sequence of amino acids. With this apparatus in place, RNA could control all the chemical versatility of proteins, and evolution could explore the possibilities of a much larger chemical space.

Günter Wächtershäuser

Günter Wächtershäuser is a German chemist and patent lawyer who has independently worked out a detailed, metabolism-first approach to the origin of life. His proposal first appeared in 1988. The scenario involves iron pyrite (a sulfur-iron mineral), serving as a catalyst in a metabolic cycle similar to one shared by all living things today. The cycle is called the citric acid or Krebs cycle—but in Wächtershäuser's account, it runs in reverse! Surprisingly, a few one-celled organisms still do this today. The reverse citric acid cycle allows these primitive cells to build the organic molecules they need from very simple inputs: water, carbon dioxide, and hydrogen.

Moreover, this cycle can grow like a nuclear chain reaction: the last step of the cycle gives us the starting ingredients all over again—but two copies of them. So long as the raw materials and co-enzyme A are available, the reaction can continue and grow. The raw materials are just hydrogen, carbon dioxide, and water, though the reaction needs co-enzyme A to be present as well, in order to complete two crucial steps. (The co-enzyme A is not consumed in the process, so it remains available for further cycles.) Co-enzyme A is a large molecule built from adenosine (one of the bases of RNA, and an important molecule for energy storage as well) along with some other components. One important feature of Wächtershäuser's account is the role of metal sulfide catalysts as precursors to modern complex organic catalysts.

Work by Michael Russell and William Martin has focused on the less extreme form of hot spring mentioned above. These springs build small chimneys as water containing dissolved minerals emerges at a (relatively mild) temperature of about 100° C. Bubbles form in the minerals that precipitate from the cooling water. Inside these bubbles, larger molecules can bind to the minerals and interact with small molecules that diffuse in from the ocean. The bubbles could serve as reaction chambers, retaining larger molecules and providing a place where they would interact to produce still more complex molecules and cycles. The boundary between the hot, basic spring water and the cold, acidic seawater would maintain a disequilibrium in which many interesting reactions could occur.

The key reaction in Russell and Martin's account is called the acetyl-CoA pathway. A primitive version of this pathway may have used iron sulfide and nickel-sulfide catalysts, minerals that form the chimneys of Russell and Martin's hot springs. The reaction uses hydrogen released from the hot springs and carbon dioxide from the ocean, producing energy-rich acetyl as waste, which is then available for further reactions. One mineral catalyst that is important in this picture is the iron-nickel sulfide mineral called *greigite*, with the formula Fe_5NiS_8. The structure of greigite closely resembles the active site of two important modern enzymes. At the right temperatures and pressures this allows it to perform the same catalytic job in Russell and Martin's primitive acetyl-CoA pathway.

Recent experiments hint at more bridges between basic organic molecules and the more sophisticated building blocks required for life. Amino acids turn out to be able to catalyze the formation of sugar molecules. A kind of bootstrap process may have emerged, whether in Russell and Martin's chimneys or elsewhere: more and more complex reactions may have built up in interlocking cycles, accumulating the molecular building blocks required for new cycles to arise, gradually laying the groundwork for the RNA world. Russell and Martin's bubbles of precipitated mineral might have helped contain the reactions' products as well as providing catalytic surfaces to help the cycles of reactions along.

Access to resources is critical to any reproductive process—without the building blocks and energy needed to keep the process going, reproduction just can't happen. So there has to be something that plays the role of metabolism from the start. At first it might take the form of materials and energy sources that were already present in the environment. Any reproducing molecules that exist in a place where such materials and energy are being produced will have a great advantage—and any that somehow manage to capture or link themselves to a reaction that produces some of the materials they need will have a still greater advantage. As variations and opportunities arise, selection will favor the most successful. Given a range of building blocks and an initial, crude capacity for reproduction, evolution could have gotten underway.

On both Wächtershäuser's and Russell and Martin's accounts it was the precursor to metabolism that produced the materials needed to get reproduction started. From there, the two developed together. On a reproduction-first model, however, the primitive reproducers developed at the outset, first drawing on raw materials available to them in the environment and only later associating with and learning to control metabolic cycles that freed them from dependence on the immediate availability of complex building blocks. But whichever came first, reproduction and metabolism have worked together from very early in life's history.

EXPECTATIONS

As we saw in chapter 4, Darwin had to explain the limits of the fossil record carefully, to show that the gaps in our record of the history of life did not undermine his evolutionary view of life. The fact that smooth transitions from fossils to today's living things are rare is still raised today as an objection to evolution. But as Darwin pointed out, the fossils we have found are a very limited sample of ancient life. The number of fossil species known today is about two hundred and fifty thousand. But there are millions of species alive today. Individual species rarely persist for more than a few million years, so, between Cambrian times and the present, hundreds of millions or even billions of species would have existed. The fossils paleontologists have found, almost 150 years after *The Origin* appeared, constitute a very small sample of past life—and the sample is biased, too. With rare exceptions our only fossils are of plants and animals with hard parts that lived in environments where their remains were buried and preserved. Even then, we can only find those that died in regions where sedimentary layers continued to be laid down. Those layers must now be exposed, or almost exposed, but not yet eroded away or subducted beneath the earth's crust. The fossils we have found probably represent less than 1 in 1,000 or even 1 in 10,000 of the species that have lived on earth. Any predictions about what we should find in the fossil record have to fit within these limits—in particular, gaps must be expected.

For the same reasons, we need to consider carefully what we should and shouldn't expect from work on abiogenesis. It is crucial that no magical vital force is required to make organic chemicals. Wohler's synthesis of urea showed that at least some organic chemicals needed no magic of life to produce them; the progress of organic chemistry since then confirms that organic chemistry is continuous with inorganic chemistry. It was also crucial that the basic molecules of life can be produced or available on the early earth—if we couldn't find a plausible source for these building blocks of life, abiogenesis would be in serious trouble.

Beyond this point, things get complicated. There is no record of the process that led to the first living things. What we have to work with along the top-down path of investigation is the biochemistry of modern life. This provides us with hints about the basic toolkits of metabolism and reproduction. Traits shared across all forms of life today, especially those that are not dictated by basic chemistry, are likely to be traits that belonged to the last common ancestor of all life. Along the bottom-up path, we have a number of ideas about early conditions and a bootstrap process leading to more and more complex, more and more stable and persistent chemical cycles and interactions. When these two paths meet in the middle, we will have a credible theory of abiogenesis.

It's clear that the chemistry required to get life started was complex and rich. But the environment of the early earth was complex and rich as well. The atmosphere, oceans and the crust itself continuously cycle in convection currents, mixing their contents as they flow, exposing new surfaces to new fluids, mixing streams of water from beneath the surface with oceans and rivers, dissolving and mixing molecules from different sources. Conditions would have varied in many ways, from temperature to chemistry to how local flows and gradients interact. The range of reactions and processes that took place in all these environments is immense. All abiogenesis requires is one group of interacting processes that got reproduction started and kept going long enough to light the fuse that burns today in every organism on earth.

8

OUR PLACE IN THE NATURAL WORLD

At last the first grand stage in the preparation of the world for man stands completed, the Oyster is done. An oyster has hardly any more reasoning power than a scientist has; and so it is reasonably certain that this one jumped to the conclusion that the nineteen-million years was a preparation for him; but that would be just like an oyster, which is the most conceited animal there is, except man … Man has been here 32,000 years. That it took a hundred million years to prepare the world for him is proof that that is what it was done for. I suppose it is. I dunno. If the Eiffel tower were now representing the world's age, the skin of paint on the pinnacle-knob at its summit would represent man's share of that age; and anybody would perceive that that skin was what the tower was built for. I reckon they would, I dunno.

Mark Twain (1903, 221)

ON THE LARGE SCALE

We are part of the natural world. Even for Linnaeus it was clear that humans have a place in the taxonomy of living things. We are linked most closely to the great apes; together with them we are linked to old world monkeys, other primates, the mammals, the land-dwelling tetrapods, and to all vertebrates. When Darwin recognized that the tree of taxonomy is really a family tree of descent, the meaning of our taxonomic place was revealed: we share a recent common ancestor with the apes, and more distant ones with the other groups listed. The variety of life on earth has been produced by descent with modification, as groups of living things became separated and altered over time by selection and drift, isolation, and environmental change. The evidence for these conclusions is as rich and strong as any in science. The distribution of life through space and time, the tree of taxonomy, and the corresponding resemblances of genes, proteins, and developmental patterns in different living things all point to the same pattern of descent.

This view of life fits elegantly into the scientific cosmology that emerged over the last century. It's worth pausing for a moment here to reflect on how much we've learned about the physical universe during that time. New, larger telescopes and improved photography revealed that there are many galaxies spread out across a far larger universe than had been suspected. The whole of that vast universe is expanding, as Edwin Hubble realized when he saw that more distant galaxies' spectra had a greater red shift than closer ones. The big-bang and steady-state models of the universe competed for decades as explanations of this fact. At the same time we were learning about the structure of atoms, which explained the periodic table of the elements, including the subtleties of their different isotopes. Studies of atomic nuclei revealed the curve of binding energy that makes fusion and fission such powerful sources of energy.

Fred Hoyle, an enthusiastic proponent of steady-state cosmology, coined the phrase 'big bang' to deride the rival theory. Ironically, his work on the formation of atomic nuclei in huge, unstable stars turned out to imply that the proportions of hydrogen, helium, and lithium in the universe matched predictions based on the big bang, confirming the theory he so despised. Hoyle also discovered how still heavier elements were gradually built up in stars. Supernova explosions complete the process, building the very heaviest elements and then spreading the matter that the earth and we ourselves are made of across space.

The big bang was confirmed in the 1960s when its afterglow, the microwave background radiation, was discovered accidentally by two scientists at Bell Laboratories. Since then, subtle variations in the early universe's density, the first seeds of galaxies and clusters of galaxies, have been detected in the background radiation. Very recently we have learned that the expansion of the universe is actually accelerating, for reasons that are not yet understood; there are, as always, more details to unravel. But the outlines of the story are well established. In particular, we know enough to understand how solar systems and planets like earth were formed.

At this point, biology comes in, and our story shifts to the origins of life on earth, discussed in our last chapter. Once again, the outlines are clear. The basic chemistry of life got underway using materials and energy on the early earth, or, just possibly, on some still earlier planet. As soon as life began, evolution began as well, eventually giving rise to the tree of life whose budding tips are today's living things.

This story of the natural world is still incomplete, and some of our questions may never be answered. We have no well-tested theory of the very earliest stages of the big bang, although possibilities are being explored. We don't know how the values of fundamental physical constants were determined. We don't have a full account of life's beginnings, or a description of all the forms life has taken in its long and complex history. For example, it's recently been estimated that fossils representing more than 70 percent of dinosaur genera, and a far larger proportion of dinosaur species, have yet to be found.

The important point here is that the existence of gaps in our present knowledge does not imply that they cannot be filled in a natural, scientific way. In fact, there are powerful reasons for thinking that this is the *only* way they can be filled. The scientific method requires us to find testable hypotheses and then make observations that can confirm or refute them. If we want an explanation that does more than just paper over the gaps, this is the only way forward. Untestable, speculative hypotheses actually leave the gaps unfilled. They only provide a kind of label for the gaps; whatever that label says, it's just an empty form of words so long as it has no links to anything else we can observe or infer. Worse, these pasted-on labels suggest that there is nothing more to know. Rather than complete our scientific picture of things, they merely put a stop to science.

Explanations that imply nothing about new observations don't lead to new discoveries—they just point back to the facts we wanted to explain. This is the same criticism that Darwin made of Richard Owen's claim, that his ur-vertebrate, the archetype on which all vertebrates are variations, was an idea in the mind of God. The biological force of Owen's suggestion did not reach beyond the facts about vertebrates that Owen used it to "explain." An explanation that leads to no new conclusions and makes no testable predictions is no better than the ancient myths that made Zeus the cause of lightning and God's anger with sinners the cause of earthquakes, volcanic eruptions, and plagues. Studying natural processes, including charge separation in storm clouds, plate tectonic movements, and the rapid spread of infections in vulnerable populations, tells us much more about these events. Our scientific ideas allow these events to be modeled, and the models can then be tested—when our models are shown to be reliable, we go on to apply them to predict and sometimes even control what happens.

We could decide to fill every gap in our knowledge of the world by invoking a miracle, large or small. But there are no logical connections between one miracle and another, or between miracles and other facts about the world. We invent these explanations case by case, each one exactly matching the gap it's meant to fill. As a result, they tell us no more than what we already knew about the world. Filling gaps in this way is useless for science, because we cannot use these explanations to extend or refine our knowledge.

A scientifically interesting way to fill a gap can't *just* fill the gap—it has to reach beyond the gap by leading us to other conclusions that we can compare with the rest of what we know, and with new observations. When Walter and Louis Alvarez proposed that a comet or asteroid struck the earth at the end of the Cretaceous, they were trying to explain the high levels of iridium in samples of rock from the boundary between Cretaceous and Tertiary. But the implications of their idea went far beyond this one fact. If the impact hypothesis was right, iridium should show up in boundary rocks from around the world, not just the few their team had examined. There should be other evidence of the impact, too. Perhaps the crater still remained, though continental drift might have buried this smoking gun under the earth's crust. Whether the crater could be found or not, other physical evidence should be found.

These predictions were soon made good. Iridium levels were high at the Cretaceous-Tertiary boundary around the world. Particles of shocked quartz produced by the impact were found in the same layer that contained the iridium. A crater of the right size and age was eventually found at Chixulub, Mexico, after evidence of a massive tsunami was discovered along the northern coast of the Gulf of Mexico. More evidence and ideas about the impact still continue to emerge. An impact striking the limestone of Chixulub would release carbonates and sulfates into the atmosphere. This may have been an important cause of the subsequent extinctions, a possibility that is being studied. Carbon from the boundary layer suggests that fragments of burning hot debris from the impact set fire to forests around the world. However, some paleontologists argue that the dinosaurs were already in decline before the impact, claiming that the variety of dinosaurs was in decline well before the end of the Cretaceous. Perhaps the impact was not the only factor in the dinosaur's extinction. How all the details fit together continues to be debated, and evidence from a wide range of sources is still being gathered and examined.

Could there still be a way to get clear evidence that something supernatural has happened, something that no natural process could possibly explain? William Dembski, a proponent of intelligent design, has tried to argue by elimination, claiming that certain facts about life are so improbable and so significant that we can rule out natural causes for them—these facts, he claims, must be the result of intelligent intervention. But his arguments fail badly, for two main reasons.

The first is that probability is a very slippery concept, and some of Dembski's claims about it are confused and misleading. Where we know the circumstances and processes involved in producing an event, we can often assign probabilities to it: if I have a symmetrical coin and toss it with a fair amount of spin, the probability of the coin landing heads is very close to a half. But when we don't know the circumstances and processes giving rise to an event, we can't assign a probability to it. This applies to the origins of life, which are not yet fully understood. Until we know the processes that took place and the circumstances these processes occurred in, we just *don't know* how probable it was that life would begin (let alone how probable any of its particular characteristics is). The probability could be anywhere from 1 (if life was certain to arise given the conditions on the early earth) to as nearly 0 as you like (if some very unlikely coincidence was required). For now, we simply do not know what this probability is—how could we, when we aren't even sure what the process that led to life involved, or whether there is just one, or a few different paths that could lead to life?

The same point applies to the origins of fundamental physical constants and other features of the early universe. We can assign probabilities to these events only by making assumptions about how they were produced. For instance, if we assume that the observed values of fundamental constants were randomly selected from all the values allowed by the laws of physics as we now understand them, then the probability of the combination we actually observe would be extremely low. But there is no reason to assume that this is how those values were settled. So we have no reason to assume that the observed constants represent a strange and improbable outcome.

Further, we've already seen that there are circumstances in which any outcome will be extremely improbable—in a fair lottery, some ticket will win even though this is a very improbable result. With just a little patience, I can easily flip a coin 1,000 times; the sequence that results will have a probability of less than 1 in 10^{600}. Any long sequence of independent natural events will have an extremely low probability, so long as each event in the sequence has a probability even a little less than 1. The mere fact that we observe something improbable does not show that we need some kind of intelligent intervention in the natural course of events to explain it. A better argument is needed.

Dembski's attempt to provide a better argument for intelligent design focuses on aims or purposes. No matter how improbable it is, a random sequence of heads or tails from my coin-flipping experiment would not count as evidence of design. But suppose I got a sequence with a striking pattern—for example, one that listed the prime numbers, beginning with a single head, then two heads, then three heads, then five, seven and so on, with each group of heads separated by a single tails? Surely this would be a strong (and startling) indication of some kind of intervention. Intuitively, the hypothesis that the sequence is random becomes less and less likely as a sequence following such a strict and clear rule gets longer and longer. However, it's important to understand why this is so.

Experience is what has taught us to seek explanations for observed regularities—we have found that, as a matter of fact, they often turn out to be due to something special about the process that gave rise to them. The idea that a sequence of heads and tails that obeys a striking and highly improbable regularity is not likely to be merely random reflects this experience. Even processes that look superficially random can sometimes be manipulated to produce non-random results. Experience tells us that such manipulations are often more likely than that a truly random sequence of heads and tails just happens to follow a simple, regular rule.

Of course we also need experience to justify adopting a particular explanation for a regularity. Still, even before we have a particular explanation, we may be quite sure there is *some* such explanation. But our confidence in this is itself a product of experience, the experience of having discovered many successful, confirmed explanations for previous regularities. The idea that there is some special explanation for the regularity is supported by our past success at finding and confirming explanations for similar regularities.

The lesson here is that before we draw the unhelpful conclusion that a new, completely unspecified and untestable explanation must apply to our sequence of coin flips, experience suggests we begin with the kinds of explanations we have already had success with, explanations we have been able to test and confirm. Ruling all these kinds of explanation out is a tall order—but only then should we conclude that an entirely new sort of explanation is needed.

Finally, when we do look for new kinds of explanation, we should begin by asking, what new kind? There is no good reason to adopt the first one that pops into our heads. Above all, we should not adopt one that cannot be tested and confirmed! This is a sure-fire dead end, a pure science-stopper.

Another problem is that by itself, a claim of intelligent intervention explains nothing. In order to do real work, intelligent intervention explanations need information about what the intelligence in question can and will do. When we have this kind of information it can be easy to recognize some products of intelligent intervention, but without it we have nothing to go on. We do know a lot about the things humans tend to do, so we can often tell that an object has been produced by human intervention. Sometimes we can do this even without knowing who did it, when they did it or why. For instance, cave paintings were always obviously human products—even though at first we had no idea of who made them, why, or when. They are recognizable images of objects and activities that were important to the humans who made them—clearly so because these objects and activities are important to many groups of humans now. They are associated with other signs of human intervention—stone lamps and other evidence that fire had been used for light, human hands outlined in paint, stone tools, and other archaeological remains. Their similarities to other paintings are so obvious that no other hypothesis about their origin is credible.

By comparison, William Paley's famous analogy between a watch found on the heath and living things is very tenuous. Living things are vastly different from mechanical devices. We know of no way to produce them other than by reproduction. So any intelligent being who created them must be vastly different from us. Paley tries valiantly to derive some conclusions about the creator from his observations of living things, but the effort fails miserably: how can Paley's creator be beneficent, when disease, misery, pain, and death are so widespread and so inevitable? What purpose can we attribute to this creator, when life's history follows no clear direction? In the course of evolution we can find both increases and decreases in complexity, both predators and prey, parasites, diseases and their victims. The world contains great beauty and true horror. On top of all that, occasionally the whole applecart is upset by mass extinctions. Of course, at the end of the process (for now, at least) we also find ourselves. But as Mark Twain pointed out, this does not show that we are the aim and purpose of it all.

Unlike the scrupulously honest Paley, proponents of intelligent design today offer no account at all of the intelligent designer's aims or means. This allows them to avoid answering these questions (or worrying about their religious implications). However, it does much more than just help them avoid some awkward questions. It isolates their so-called theory completely. Without some information about the designer's aims and means, we have no idea what to infer from the hypothesis that there is a designer. The content of this empty intelligent design hypothesis is indistinguishable from the more modest claim that no *known* process has produced life as we know it. If it succeeded, Dembski's argument by elimination would show only this: that we do not know how life arose.

But in fact the argument by elimination fails. Evolution is directly supported by powerful evidence. First, the evidence powerfully confirms common descent. Detailed homologies of anatomy, physiology, and development parallel the basic chemical resemblances of proteins and genes that living things share. Nothing in our understanding of chemistry and physics suggests that life's beginnings or its evolution require some kind of special, magical intervention.

Natural selection ensures that any heritable change that gives a consistent overall advantage to its possessors tends to spread through a population. Of course natural selection does not predict particular changes in advance. It's also very difficult to predict whether a given mutation will be advantageous, since it is generally impossible to anticipate all of its effects. But natural selection still passes real empirical tests. We can predict that genes that produce resistance to a disease will be common where that disease is common. The sickle-cell anemia gene is a perfect example: it confers resistance to malaria, and it is common in populations that come from regions where malaria is a major health risk. In those regions, the fact that anyone unfortunate enough to carry two copies of the gene will die of anemia was compensated by the fact that those who carry just one sickle-cell gene have a better chance of surviving malaria. As a result, the gene persisted in those populations. We can also predict that disease organisms will evolve in ways that reflect the success of different strategies in the overall population. Cholera bacteria are an example of this. Cholera can be very virulent, causing serious illness or death, or it can be relatively mild. Where drinking water is contaminated by sewage, the more virulent form spreads very quickly because victims who are seriously ill develop severe diarrhea and release huge numbers of the bacteria that infect new victims who drink the contaminated water. However, when drinking water is safe from contamination, milder forms of the bacteria do better. Their fairly healthy victims continue to have social contact with other people, providing more opportunities for the milder bacteria to spread. One last example here: Sewall Wright argued, in his account of adaptive radiations, that a species which develops a new and highly advantageous trait tends to spread widely and give rise to many new species. These develop variations on this trait, adding new features and specializing in ways that take advantage of the trait. This process provides a good general explanation of a recurring pattern of evolution, revealed in the fossil record of groups like trilobites, dinosaurs, birds, and mammals: each of these groups grew from small beginnings, and then spread widely, developing into many different forms.

Everything we know about biology supports the evolutionary account of life's history. As Darwin pointed out long ago, biology repeatedly creates new adaptations by tinkering with what is already at hand, producing variations on established themes and building new organs out of old: lungs developed out of swim bladders, bird and bat and pterosaur wings out of forelimbs, and the intricate structures of orchid flowers developed out of already existing flower parts. Organs and traits do not appear suddenly and without precedents—but if life was designed by an intelligence, there is no reason why entirely new organs should not have been simply added to organisms that need them, rather than developed as sometimes awkwardly altered versions of earlier organs.

Plausible, useful intermediates exist even for complex structures like eyes: we know of no complex, useful trait which is *isolated,* that is, not part of a series of increasingly simple but still useful traits out of which it could have evolved. It's important to recall here that the uses of simpler forms of the trait often differ from the current use of the complex form: nature is no respecter of single, fixed functions. Birds do not hesitate to use their wings to shelter chicks or drive off predators. Teeth can be very useful for holding things, even if their main function is for eating: when I was very young my parents tested a new kind of child-proof bottle on me by handing me the bottle and asking me to open it. I couldn't do it with my hands, so I used my teeth—no problem. Even at the biochemical level, multiple effects are common. New proteins selected for one chemical trick are later recruited to play other roles (and then refined to fit them better by natural selection).

THE TROUBLE WITH MIRACLES

The hypothesis that *God did it* is not logically contradictory. In principle, some of the facts we can't yet explain naturally could be the result of direct intervention by God. But the hypothesis contributes nothing to science because it provides no basis for further investigation. Worse, history shows that explaining things we can't (yet) explain naturally by invoking God is a losing strategy. Long ago, gods were thought to be directly responsible for the operations of the natural world. Storms and winds, the motions of the heavens, earthquakes and volcanic eruptions, even plagues and diseases were seen as their doing. The monotheistic God was once thought to intervene directly too, moving the planets in their orbits around the earth, punishing people and nations with disease and disaster, warning us with comets and other prodigies, and so on. But we now have natural explanations for these events—explanations that often allow us to predict them, to identify and explain many more facts about them, and sometimes even to avoid or prevent them.

When someone uses God to fill gaps in our natural explanations and then uses these explanations as evidence for God's existence, things get even worse. Such arguments for God's existence depend on the continued failure of natural science to fill the gaps. This creates unnecessary conflict between science and religion. Gaps are always in the process of being filled. As science advances, more and more of what we know about the natural world is explained. But these new natural explanations, no matter how well-tested and confirmed, are resisted by many religious believers because the explanations compete with their gap-filling ideas about God's role in the world. Science's advances become religion's retreat, and defenders of religion come to see science as a rival and enemy.

For the purposes of science, nature is *explanatorily closed:* natural events are assumed to have natural explanations. The most important characteristic of natural explanations is that they rely on rules and patterns which connect them to what happens in other cases. Supernatural explanations break these rules, because they aren't subject to the laws that connect natural events. This does make supernatural explanations easy to give, but it also makes them arbitrary and uninformative. Supernatural explanations are limited to the specific case or phenomenon they are aimed at, because they don't obey general rules that we could use for prediction and testing.

If we do generalize a supernatural explanation by applying rules and principles so that we can draw new conclusions and predictions from it, then the result is really a new, natural theory. Once there are rules and patterns to test, the scientific content of the theory just is those rules and patterns. Their association with a supernatural entity is not scientifically important, because it does no work. For example, even when different nature-gods were associated with storms and winds, many features and patterns of these natural phenomena were known. But this practical knowledge of natural operations and their regularities did not depend on their theological associations; when belief in these gods waned and disappeared, the familiar facts remained—after all, the natural patterns had been doing the real work from the beginning.

Some modest theologians even see the explanatory closure of nature as part of the perfection of creation: a creation that is complete in its own terms has an elegance and wholeness that an incomplete, dependent creation would lack. But the price of their modesty is that they must give up natural theology. If nature is explanatorily closed, nothing in the natural world points *outside* the world.

Charles Lyell rejected catastrophic hypotheses in geology because they were very difficult to investigate, positing processes that we could neither observe in operation today nor model theoretically. He replaced them with familiar kinds of events taking place over immense periods of time. These events could be studied, their effects and the traces they leave identified and compared against the geological record. In case after case, Lyell's uniformitarian method was applied with great success.

Similarly, explanatory naturalism rejects miraculous interventions, large or small, because they are *impossible* to investigate. It replaces piecemeal gap-filling with an ongoing search for a natural, scientific history. Our continued success in producing new, testable natural explanations confirms the value of this approach.

Some people are reluctant to accept such a retreat for religion. Traditional religious beliefs have always included mythological elements meant to explain aspects of the natural world. Religion has been invoked to explain many events—disease and recovery, war and peace, bad harvests and good are often attributed to God, whether as direct punishments and rewards for human behavior or more modestly as part of a complex plan we don't fully understand. But the explanatory closure of nature excludes such stories from science, directing scientists to seek explanations in the workings of nature instead. This is not because these religious stories are nonsensical or contradictory. It's because they close off investigation while adding nothing to our ability to make reliable observations and inferences. We have found many natural, scientific explanations that pass demanding observational tests and contribute to further successful investigations. This is what makes the pursuit of natural explanations *objective*, rather than a matter of personal faith. Scientific explanations are not just tailored to fit facts that we already know. They continue to be tested as new predictions and observations are made, and they succeed only so long as they go on leading us to new facts we don't already know. This doesn't make our present scientific view of the world the last word—but it does make it part of a successful process in which answers to more and more questions are tested and *improved* over time.

ON EVIDENCE AND RELIGION

The explanatory closure of nature separates religious beliefs from scientific ones. By studying science and its methods we can come to understand the standards that scientific beliefs have to meet to be justified. This leaves the question of whether and how religious beliefs might be justified open.

Some philosophers have argued that religious beliefs can be justified in the same way simple observations are. Simple observations don't seem to need support from other beliefs to be justified: just by looking at something I can arrive at a justified belief that it is red. Similarly, religious beliefs may not require support from some other kind of evidence to be justified. This makes the justification of religious beliefs a matter of personal conviction rather than shared, public evidence, because justification for these basic beliefs depends on private, personal experience. There are difficulties for this kind of account, but it does at least allow for a separation of science from religious belief.

However, many religious believers reject scientific explanations of life, the universe, everything. A crucial bone of contention for these believers is the place of human beings in nature. For them, the claim that we descended from a recent common ancestor with the apes, and from more remote common ancestors shared with all mammals, all vertebrates, and even all life on earth, is unacceptable. This raises a couple of basic questions: why do so many people continue to reject a scientific account of our place in the natural world? What beliefs does it threaten, and why are they so important to so many?

Some claim that evolution undermines morality. The crudest form of this argument tries to blame evolution for terrible historical crimes. For example, the Nazi party claimed that the so-called German race was the most evolutionarily advanced. This superiority was supposed to justify fighting a brutal war of aggression to achieve world dominance and eliminate any threats to their power. However, the Nazis also appealed to a long and shameful tradition of Christian anti-Semitism for support. So if evolution is to be blamed for its small contribution to Nazi rhetoric, Christianity will have to face the same charges.

More importantly, it's just plain silly to blame a scientific theory for the use that a vicious political movement has made of it. There is no scientific support for the existence of a distinct German race, and there is no evidence that Germans are genetically superior to other humans. Further, even if these absurd claims were true it wouldn't follow that a German race would have the right to dominate the world and exterminate its rivals. Nazi propaganda used evolutionary *talk* to persuade some people that their paranoia, racism, and aggression were justified. It also used religious *talk* for the same purpose. This doesn't mean there's something evil about either evolution or religion.

Another attempt to cast evolution in a bad moral light argues that, if we are products of evolution, then no real ethical standards apply to us. The gist of the point can be conveyed by asking some questions: if we humans are really just a peculiar kind of ape, what do *right* and *wrong* mean to us? Aren't animals free to do whatever their impulses tell them to? If we're just animals descended from other animals, aren't we free to act as we please, too? Of course questions like these don't constitute an argument. No real argument follows from their answers either. By itself, the fact that we have evolved from other animals has no ethical implications at all. It's true that we don't apply ethical standards to the actions of other animals. But we do apply them to ourselves and other human beings. We have evolved from animals that we wouldn't apply ethical standards to. But this only implies that we now have some properties relevant to ethical standards that our distant ancestors lacked. This isn't even strange or surprising; after all, the same goes for each of us individually: we don't apply ethical standards to the behavior of infants. But we do apply them to the behavior of adults. It follows that adults have some properties relevant to ethical standards that infants lack. But of course we knew this already.

A third argument claims that evolution does provide moral standards, but they are bad standards. Proponents of this argument ask, why should we care about others? After all, evolution favors only those who manage to out-survive and out-reproduce their competition. They conclude that, if we are the product of evolution, we should be ruthlessly selfish; generosity, kindness, and (above all) self-sacrifice make no sense at all. If the argument succeeds, an ethics based on evolution must be a harsh creed of survival of the fittest and devil take the hindmost.

This argument mistakes a caricature of evolution for an ethical theory. *Homo sapiens* is a very social species. A solitary human is a rare and unhealthy animal. Empathy, the ability to understand how others feel, is a critical skill for us—it helps us to manage our social relations, avoid conflicts, and maintain cooperation. Sympathy for others is also normal for human beings—as it is for many other species as well. Evolution may have favored such emotions because they helped to maintain group ties, or because they made us more likely to help family members, or for any number of other reasons. But the conclusion is the same regardless: an ethics based on our evolutionary heritage would not support ruthless selfishness at all.

Human societies around the world value generosity, fairness, and concern for others. Individuals are capable of selfishness and aggression, but we also share, help others, and protect the weak. These positive virtues are as firmly grounded in our evolutionary history as any human vice. Societies around the world teach them to their children successfully.

More fundamentally, basing our ethics on evolution is wrong-headed in the first place. Evolutionary history does not impose sharp limits on what we can do or how we can think. We often use our minds and bodies for different purposes than those that shaped them in the course of evolution. Our hands were not shaped by selection for the ability to type—but they are pretty good at typing all the same. At least until very recently, our minds and brains were not shaped by selection for thinking in terms of ethical principles. But as intelligent social beings, ethical ideas and commitments have become very important to us. Even if our capacity for sympathy is based in an evolutionary history of kin-selection, humans still feel sympathy for non-relatives. Even if the virtue of generosity has its evolutionary roots in the fact that being generous helps us climb social hierarchies, many people still choose to give anonymously. Evolutionary history does not dictate what's ethically appropriate for us, or how we should use the capacities our history has given us. Ethical judgments are grounded in much more than our evolutionary roots—they are grounded in ethical thinking and ethical instruction; in caring about each other's needs and concerns; in ideals like fairness, duty, and obligation; and in values like well-being and happiness.

DESCRIPTION AND NORMS

Evolution is a part of descriptive science. It aims to explain how life has changed through time, how different groups arose, spread, and died out, how life today has descended from ancient forms of life. But it does not, and cannot, tell us whether what has happened was good or bad. Neither can it tell us what we should or should not do. Dinosaurs died out at the end of the Cretaceous, along with many other groups. If they hadn't, it's unlikely that humans would ever have evolved. So we might be inclined to regard the mass extinction as a good thing. But it was also a horrific disaster for millions of living things around the world. Many died of terrible burns, drowning, or being brutally crushed; others suffered slow, agonizing deaths from cruel injuries, poisoning, and starvation.

The story of evolution does not *evaluate* any of this. It only describes the history of life and the processes involved in it. Nevertheless, from its earliest proponents to today, many have been tempted to identify evolution with *progress.* This mistake contributed to the development of *social Darwinism,* a harsh political ideology that rejected humanitarian intervention in the economy because helping the poor would reward the so-called unfit at the expense of the fit. Herbert Spencer, who coined the phrase *survival of the fittest,* was the founder of this misguided view of evolution. The famous Standard Oil tycoon, John D. Rockefeller, was an enthusiastic supporter of Spencer's views. According to social Darwinism, Rockefeller's success at undermining, ruining, and buying up his competition was proof that he and his company were fit and *deserved* to survive and prosper. Social Darwinists often advocated ruthless and aggressive competition—a stand that Rockefeller, whose manipulations and double-dealing were legendary, could truly appreciate: where else could he find a view of ethics that praised his success and power, however they were gained, as proof of his evolutionary virtues?

But evolution is not progress. Progress is supposed to be good, a process that moves us towards some goal that we value. Standard Oil's monopoly control of the oil industry served Standard Oil's interests, not the economy's or other people's. The manipulations and schemes that propelled Rockefeller's success did not ensure an improvement in human welfare. They just made

Rockefeller's company richer. Evolution is a natural process. It is not aimed at any particular goal, let alone one that we value. Sometimes evolution produces greater complexity and sophistication, but sometimes it simplifies things. The evolution of parasites often eliminates traits needed for an independent life, producing a streamlined, more efficient parasite. In different circumstances, we might evolve smaller brains (after all, our large brain demands a disproportionate share of the body's resources, while also making childbirth difficult and dangerous for both mother and child). Evolution has led to many things we value greatly, including our own existence, but it has also produced diseases, parasites, and many other things that threaten what we value.

It may still be tempting to think that the process that gave rise to us must be a *good thing*. But evolution does not demand our endorsement or support—and often does not merit them. In an environment without good sewers, evolution can produce a deadly, fast spreading strain of cholera. We rightly regard this as a bad thing, just as we regard preventing evolution from taking that course by building better sewage systems and reliable sources of pure water as a very good thing. But either way, evolution itself is like gravity: neither good nor bad.

To think about ethics and morality, we need to bring our own values to bear, not blindly cheer for more evolution no matter where it leads. Concern for human welfare, for good health, nutrition, and shelter, for access to basic opportunities in life: these are important general values. Suffering, both human and non-human, is generally a bad thing, although no doubt some level of suffering is inevitable, and we accept some suffering because it is inseparable from good consequences that outweigh the suffering.

Our goals are not the same as evolution's pseudo-goal of survival and reproduction. In us, the blind workings of evolution have produced an organism that really does pursue goals. Evolution by natural selection works because inheritable traits that contribute to survival and reproduction tend to persist and spread, while those that detract tend to disappear. This is all there is to natural selection. Evolution does not look ahead to anticipate what might work in the future—but human beings do. Natural selection only shifts direction when the environment has already changed. Whatever the situation, natural selection favors whatever traits contribute to survival and reproduction in that situation. But human beings can anticipate and prepare for new situations. We can shift direction before the environment does. Like some other animals, we build shelters and lay food aside for winter. But we take this much further. We innovate and learn how to build better shelters. We don't just set aside food that keeps well, we invent techniques and machines to preserve food that would otherwise rot. We also teach these techniques to others, and we study, discuss, and improve on them. Our goals really do concern the future; when our plans fail, we've made a *mistake*. We may even be able to stop the world from changing in ways that would threaten us and other things we value. By contrast, evolution simply *happens;* no mistake is involved when a lineage goes extinct. Evolution has no plan for the future and it seeks no goals.

PURPOSES IN THE NATURAL WORLD

Though evolution does not undermine ethics, there is still a difficult philosophical question here: what is the place of *purpose* in the world? In chapter 1, we saw that Aristotle took purposes, in the form of *final causes*, to be part of the natural world, working right alongside the more familiar *efficient causes*. But modern physics has turned away from final causes—purposes or goals that a system tries to reach no longer play a role in physical explanation. The forces at work in a system arise from its state here and now. They explain what it does without appealing to special states that it tends towards or aims at. Even when a system does tend towards and eventually reach some state, this is explained by efficient causal processes. The end state does not explain what the system does; instead, the end state is explained by how the system *works*.

Evolution by natural selection has extended this causal view of the physical world to living things. Even though they have complex adaptations that enable them to find nourishment, grow, and reproduce, living things can be understood without invoking final causes. Evolution explains the *appearance* of final causation in the living world without assuming any built-in purposes at all. Any system that is better than its fellows at surviving and reproducing will tend to produce more copies of itself than its fellows do. If its descendants continue to have an advantage, they will eventually dominate the population, even if they began as a very small group. In fact, if there is competition for resources, the other forms are likely to become extinct. As Darwin argued, so long as new heritable variations continue to arise and some of them provide an advantage, natural selection will continue.

This process can build very complex structures. To understand how this is possible, it helps to recall that variation comes in many forms. The simplest kind is small changes to individual genes. Natural selection applied to such changes can refine the genes' contribution to a particular, fixed effect. It can also select for a new effect produced by a variant form. Variation also occurs when entire genes and sets of genes are duplicated. This may have no dramatic effects on fitness at all, at first. But after duplication, the two copies can change, each becoming adapted to different functions. Sometimes the two become mutually dependent, each unable to do everything that the original gene did, but together doing a better job of it. Sometimes parts serving one purpose come to serve another: precursors of insect wings may have helped insects to warm up in the morning sun—and even a slight increase in the insect's size could make these solar heaters into aerodynamically effective gliding wings. As soon as a new use begins to contribute to survival and reproduction, new variations can be selected for. Wings used for gliding can gradually become wings for active flying, as increasing levels of strength and control are selected for.

As we saw in chapter 6, patterns of expression for genes can also change, altering how the organism develops. The basic vertebrate plan of a nerve cord running along the back and a gut running along the front of an animal apparently arose from an animal with several nerve cords running along its exterior and a gut running down its middle. Building on that original architecture, mollusks developed a different arrangement from our own, with a nerve cord running along the front and a gut running along the back. The very similar proteins that guide development in today's vertebrates and mollusks point to their common ancestor's more general developmental plan. Geoffroy's idea that these two groups shared basic structural features despite their differences turns out to have been right.

Darwin's theory explains these changes without invoking goals or purposes. Differential success at reproduction due to inheritable traits is all that's needed. What remains of Aristotle's teleology today is no more than a delicate analogy. We can truly say (for instance) that the human heart has been shaped by selection for its ability to pump blood through our arteries and veins. So it may seem that we can explain the way the heart is by appeal to what it *does* for us, that is, by appeal to its purpose. But a more accurate view of natural selection's contribution focuses on what it *did,* that is, on how it contributed to our ancestors' survival and reproduction in the past. Natural selection looks to the past to explain the present, just as modern physics does. But truly teleological explanations appeal to the future, to a *not yet achieved* goal or purpose, to explain what happens on the way to that goal.

Goals or purposes no longer have any place in our scientific understanding of the world. Gravity shows no partiality to us and our purposes, or to any other purposes either—it operates in the same way regardless of whether the effects are good or bad, whether a house is being held solidly on its foundation or a child is falling out a window. The same goes for all the basic operations of nature. Chemistry goes on following its laws whether the effect is to nourish someone or to poison them. The laws make no reference to goals or purposes at all. Considered from the point of view of pure scientific description, the world just is the way it is, and things in the world simply do what they do.

I think that this is why so many people see science's world view as cold, impersonal, and amoral. The scientific description of the world makes no mention of purposes or values, and it's easy to mistake this absence of purpose and value in our scientific descriptions for an *incompatibility* between the world as science describes it and the reality of purposes and values.

However, unlike its description of the world, science itself is full of goals and values, just like any other human endeavor. There are goals and values that characterize what scientists seek to achieve: a well-tested, powerful understanding of the natural world. Science also aims at many more specific goals and values: to explain interesting phenomena, to propose and test new theories, and to make new and better observations. Scientists in training learn to present arguments drawing on theory and observations; they learn standard lab procedures that ensure reliable, reproducible results. These scientific values are not just arbitrary elements in the so-called game of science: they are part of a general effort to refine and improve on our commonsense understanding of the world—to find descriptions of things that are more reliable, more informative, more detailed, and better tested and supported by evidence.

Furthermore, there is no incompatibility between these scientific values and other human values. In fact, our desires for security, good food, comfort, social opportunities, health, and more can all be pursued more successfully when we have a good scientific understanding of how they might be achieved, as well as what might interfere with our efforts to achieve them. While the scientific description of the world does not speak of purposes or values, it is an essential part of how we pursue the purposes and values we have. Despite the fact that it leaves values out of its description of the world, science does not rule values out—scientific descriptions of the world are powerful tools that help us to achieve our goals, tools that have become essential to us as our dependence on technology has grown.

THE NUB

Explanatory naturalism rules out any role for the supernatural in science. Science aims to be a closed system, appealing to natural facts and principles to explain natural facts. Beyond such facts and principles, it has nothing to say about the world. But goals and purposes don't appear in scientific descriptions of the world—none of physics, chemistry, or evolutionary biology make any use of them. This leads to a difficult question: if the scientific view of the world leaves out goals or purposes, what makes us think there really are goals or purposes at all?

A direct answer is to say that we cannot understand ourselves at all apart from purposes and goals. Thinking in terms of purposes and goals is indispensable to us. We understand ourselves and each other as goal-seekers. We deliberately choose to do (or not to do) things because we see that they are valuable in themselves, or that they lead to some end we value. We are future-oriented—when I plant tomato seeds in the spring, it's because I hope and expect them to produce tomatoes later in the summer. If they don't, something has gone *wrong*. Purposive thinking is future-directed in a way that natural causes are not. This kind of deliberate, goal-directed action is the basic model for teleological thinking in general.

Aristotle's natural teleology applied a modified sense of the idea of deliberate action directed towards an end to some features of nature, especially living things. Paley's natural theology made this parallel literal, interpreting living things as products of God's deliberate, conscious design. But modern science has undermined the parallel, reducing it to a weak analogy. Wallace pointed out that Darwin's phrase *natural selection* suggested a stronger parallel than he and Darwin actually believed in. Natural selection is *not* directed towards a future goal—like other natural processes, it is a causal consequence of the local facts. All that matters is which organisms survive to reproduce and which do not. The analogy is that, so long as some heritable features consistently make a contribution to survival and reproduction, they tend to become more common over time. As a result, when conditions are stable enough, selection can produce refined and complex adaptations that enable organisms to survive and reproduce successfully. In some ways, these adaptations look *as if* they are specifically directed towards these general goals, and towards the specific effects by which they contribute to them. But they are not.

This leaves us with a question: what is the relation between the fact that we think of ourselves and what we do in terms of goals, purposes, deliberation, and intentions, and the purely descriptive, scientific account we can give of ourselves as part of the natural world? Here we find reasons for wondering whether we can regard ourselves as *just* another part of nature. Can we combine our commitment to understanding ourselves in terms of goals and values with the idea that we are natural beings living in a natural world? If a natural world, including us, can be fully described without reference to goals and values, these two views of ourselves may be irreconcilable.

The idea that our teleological understanding of ourselves is incompatible with a purely natural description of what we are rests on two assumptions. The first is that the descriptive story science aims to tell about the world is complete, leaving nothing out. The second is that this descriptive story does not include real goals and purposes, but only the weak analogies to them that emerge from natural selection.

When we considered what science is about, we concluded that our scientific description of the world is itself aimed at certain purposes. It provides an account of the world that explains how things are by appeal to natural processes and the rules that govern them—an account we test and improve over time by careful reasoning and systematic observations. There are good reasons to think that having an account like this is an important goal—we need reliable information about how things are and how they work in order to decide what we should do to achieve the things we value. Scientific investigation is a powerful source of reliable information about such facts. Important as they are to us, though, the descriptions science gives us of the world say nothing about goals, purposes, and values. These are *normative* ideas, ideas about how things should be, what we believe is worth pursuing, what we want to achieve. If we ourselves are natural things, we need to reconcile the fact that we apply these normative ideas to ourselves with the fact that a purely scientific description of us, and of the world, seems to say nothing at all about them.

Philosophers have proposed four general ways of dealing with this tension: elimination, reduction, dualism, and compatibilism. Elimination concedes the point, and proposes that we give up on normative talk altogether. If the world is just as science describes it, and science includes no norms or values, then norms and values are some kind of illusion. This radical stance seems to have the advantage of economy. Occam's razor—the principle that, other things being equal, we should prefer a simple hypothesis over a more complex one favors it, because it postulates less than the other three positions. But using Occam's razor well requires great care. It is a bad idea to add superfluous things to our account of what there is. But it's also a bad idea to do away with things that are important. Elimination directly contradicts our normal understanding of everything we do. Rather than aim at anything at all, elimination describes us as passively responding to the causes that act on us, just like everything else in nature. Eliminativists are prepared to bite this bullet. They claim that an objective view of what we are must dispense with talk of norms and goals. For them, this apparently indispensable way of understanding ourselves is just a collective illusion.

Reductionism is less extreme. It claims that values and goals are really there in the scientific description, but they aren't *labeled* as such. According to reductionism, somewhere in our evolutionary history we developed the ability to have real preferences, goals, and values. We became truly teleological, end-directed things, even though the natural world from which we developed is not teleological itself. We can see, in outline, how this story must go. Natural selection favors any heritable change that improves the odds of an organism's surviving and reproducing. Complex multi-cellular animals like human beings have pursued a very demanding strategy in the evolutionary game. We use our senses to acquire information about our environment. This information is processed in our brains, which then guide our actions. Most brains are complex enough to learn at a basic level, developing new associations in response to the results of previous actions. Sophisticated brains can process information in very rich ways, coordinating complex and adaptable social behavior and, in the case of humans, sophisticated tool-making and language use. According to the reductionists there must be some point in this process where we not only behave *as if* we had goals and values, we really do have those goals and values. The weak analogy to teleology that applies to evolution in general has been transformed into real teleology in us.

A vivid illustration of the need to draw this line somewhere can be seen in the behavior of a famous group of parasitic wasps of the genus *Sphex*. These wasps prey on insects and spiders, paralyzing them with venom and then dragging them down into a burrow as food for their larvae. As we watch a wasp at work, it seems obvious that it's pursuing a goal: to get its prey down into the burrow to feed its offspring. But the wasp's behavior is startlingly rigid and inflexible. Interfering in a very small way traps the wasp in an aimless, never-ending behavioral loop.

Once the wasp has dragged its victim to the brink of the burrow, it goes down into the burrow, apparently to check on whether everything is alright before pulling the spider down. Normally the prey is still right at the burrow's mouth when the wasp comes up again. The wasp grabs it, drags it down, and lays the eggs. But if we move the prey just a few inches from the mouth of the burrow, things go off the rails. When the wasp comes up again it finds its prey nearby, pulls it to the brink of the burrow, and once again leaves it there to check below. The wasp will repeat this cycle so long as we continue to move the wasp's victim a short distance from the burrow while the wasp is down in the burrow. The wasp never decides, "Well, I've checked a few times already, so this time I'll just drag the spider down without checking again." It automatically checks the burrow every time it manages to get its prey to the burrow's lip. The complex behavior that seemed so obviously aimed at providing food for its offspring is the product of a simple, rigid routine, and the wasp is totally insensitive to the fact that the routine is no longer working. This behavior is so mechanical that it seems just plain wrong to claim that the wasp is really pursuing the goal of providing for its offspring. Instead, it seems to have a simple behavioral program that usually leads to that result—and of course, that is all that success in the so-called evolutionary game really requires!

We might wonder, in a modest spirit, whether at some level of description human beings also behave in similarly rigid and unproductive ways. Examples may even come to mind. But we know that humans would quickly notice any such obvious interference with their aims, and they would respond by changing their behavior. We are not rigidly programmed machines, and we are often, if not always, sensitive to whether what we're doing actually works! Our behavior is responsive and flexible enough that it seems right to say that we really do pursue goals.

The reductionist still faces a difficult problem, though. It's difficult to decide exactly when an organism's behavior has become so adaptive and flexible that it really is goal-seeking, rather than merely simulating goal-seeking. Evolution has produced animals whose behavior ranges from extremely basic patterns of stimulus-response, through rigid routines like that of the *Sphex* wasps, all the way to the complex, adaptable behavior of human beings. Choosing the exact point in this range where real teleology arises seems arbitrary at best, and absurd at worst: how could a small difference in flexibility and adaptability suddenly give rise to a sharp and very important difference in kind, moving us from the mere illusion of teleology to the real thing?

The dualist approach declares that teleological beings actually have two separate (though closely related) parts. One part is a natural thing that science can describe completely. The other part is where real values, goals, purposes, and so forth reside. This point of view automatically generates the nice, sharp line that is so difficult for reductionists to draw. The natural beings that have this non-natural part are truly teleological, while those that lack it merely simulate teleology.

But the advantage of having a theory that can draw a sharp line is offset by the problem of deciding where individual cases fall. The line-drawing problem doesn't really go away. The dualist *assumes* that there is a correct answer about where to draw the line, based on her dualist metaphysics. But this doesn't help us to draw the line. Worse, it turns the source and reasons for the existence of the division into a mystery that science has no way to illuminate. The non-natural, truly teleological part of us is separate from the gradual development of increasingly complex and subtle behavior that took place as we evolved. Somewhere along the way, the dualist claims, something was added to this natural development, something distinct from all of the objects and processes of the natural world that science investigates. This is a *very* strange claim. It needs both clarification—just what was added and what difference does it make?—and justification—how can we detect this difference?

The last philosophical approach to this question is compatibilism. This approach begins by noticing a difference between two kinds of *language*, aimed at different purposes. *Descriptive* language, refined over time with the progress of science, aims to describe the entire world, including us as part of that world. But *normative* language interprets us and what we do in terms of values, goals, and purposes.

Good descriptions must fit the world as we observe it to be. They also need to connect those observations together, allowing us to infer how things develop and change over time. Doing both these things is very difficult, since our observations are generally independent of our inferences. Halley famously inferred that if his comet was on an elliptical orbit, it would re-appear in that orbit about 76 years later. But the observation of the comet's return long after Halley's death did not rely on Halley's reasoning—it relied on human eyes and telescopes, many of which independently confirmed that his inference had been correct. These are the standards that science is held to, standards it has been astoundingly successful at meeting.

Good interpretations are measured by a different (and more controversial) standard. They are evaluated by whether or not they help us think about actions successfully. Good interpretations should help us function as individuals and in our social interactions. But very different interpretations can all be reasonably successful, a fact that makes it difficult to compare and improve our interpretations. Worse, when I interpret myself or someone else as seeking some goal—to find some food, for example—this doesn't predict what they will do very well. Even if there is some food in the trunk of my car, I won't go and get it unless I'm aware that it's there. In fact, even then I might prefer to seek food elsewhere. Still, if my eyes are scanning the road as I drive, if I'm spending more time looking at restaurant signs than any others in view, and if I finally do park near a restaurant, get out of the car, and walk in, the interpretation that I was seeking food fits the facts pretty well. The interpretation's success is further confirmed when I give this as the reason for my behavior.

It is important to note that we can misinterpret even our own behavior, as some striking experiments have shown. One such experiment used electrodes to trigger arm motions. But when asked why they had moved their arms, the subjects gave reasons for the motion, such as "I was waving," or "my arm was feeling uncomfortable in that position." The subjects thought their movements were deliberate and motivated, even though the experimenters had directly caused them. We sometimes seem to detect intention when it's not really there—even in our own case. But we can learn to be modest, too; we can learn to test our interpretations and evaluate them more carefully. A little more information and testing might allow the subjects of such experiments to realize that their odd arm motions were not deliberate or intentional at all.

The interpretations we've been talking about are part of a system of ideas we apply to ourselves and others. Learning to interpret what others are aiming at and to predict what they are likely to do is essential to social primates like us. Though we know that people don't always do what they should, we also know that their actions usually do make some kind of sense. In fact, the ways in which we normally describe what they do are tailored to fit into this kind of explanation: we don't describe behavior in terms of specific motions or muscle contractions.

We describe behavior in purpose-oriented ways: reaching for something, walking somewhere, tasting something, and so on. Teleology is part and parcel of almost everything we ordinarily say about our own actions and those of others. The active engagement with the world expressed in this kind of talk is radically different from the passive, purely descriptive accounts we find in science.

For the compatibilist both descriptive talk and teleology are indispensable aspects of how we understand ourselves and the world. The success of science has provided us with a powerful explanatory account of how things stand in the world. But we need to combine this with an account of what we seek and what we value in that world, an account that is teleological from the ground up. Compatibilism doesn't treat the relation between scientific description and goals and values as a metaphysical issue. Goals and values are part of a way of thinking and talking about the world that is *justified*, but by different standards than those that justify scientific descriptions. There is no need to reduce talk of goals and values to descriptive facts, and no need to posit non-natural things that make claims about goals and values true. The temptation to read the differences between these languages in metaphysical terms is hard to resist, but compatibilism asks us to think instead of sentences and theories meeting the standards that apply to them, that is, as serving some purposes.

We needn't choose between these four philosophical theories here. What is important for us is this: only eliminationism rejects the existence of purposes and goals. By doing so, it cuts us off from our normal understanding of ourselves and our actions. But the other three views accept the evolutionary story of life and human origins while preserving a place for teleology in our account of the world. This is enough to show that the idea that we really do have purposes, and that our lives therefore have meaning for us, is compatible with the idea that we are natural things in a natural world, the products of a long evolutionary history.

A FINAL WORD

Like other descriptive facts about the world around us, the facts of evolution are important guides to action. Medical research often draws on evolutionary principles. Agriculture has relied on them too, at first unconsciously, as we shaped wild plants and animals into crops and livestock by steady selection, and today quite consciously, as we seek genes in related wild stocks that could provide useful new traits such as disease resistance, and as we consider the evolutionary implications of our attempts to control pests in different ways. Evolution illuminates the facts of genetics and explains why the study of other living things has taught us so much about ourselves, as well as why some living things can teach us more than others. We still have much to learn about the long process that produced the wonderful richness and variety of life on earth. But we have learned the central, unifying fact of biology: all living things are related. Together with all our earthly cousins, distant and near, we have a history and a world in common.

TIMELINE

347–343 B.C.E.	Aristotle writes influential works on biology.
1665 C.E.	Athansius Kircher publishes *Mundus Subterraneus* (underground world). Regards some fossils as remains of animals, turned to stone, but also regards some as the product of a seminal aura or plastic virtue.
1665	Robert Hooke first uses the word *cells* to describe the microscopic units of life.
1668	Francisco Redi conducts experiments undermining spontaneous generation.
1669	Steno's *Prodromus* proposes three principles of stratigraphy—superposition, original horizontality, original/lateral continuity. Defends the view that fossils are the remains of living things.
1686	John Ray publishes *Historia Plantarum*, using a taxonomy based on observed patterns of similarity and difference.
1735	Carolus Linnaeus publishes *Systema Naturae*. First use of binomial nomenclature (Genus, species); names the basic taxonomic ranks: species, genus, family, order, class, kingdom.
1753	Linnaeus publishes *Species Plantarum*.
1765	Lazarro Spallanzani conducts experiments undermining spontaneous generation using sterilized, sealed jars.
1774	Abraham Gottlob Werner presents a system of mineralogy in *On the External Characters of Fossils, or of Minerals*. He becomes the leading figure of the Neptunist school of geology.
1794–1796	Erasmus Darwin advocates evolution in his *Zoonomia*.
1795	James Hutton publishes his *Theory of the Earth*.

1796	Georges Cuvier 1769–1832. Cuvier publishes a paper on fossil elephants, convincingly establishing extinction.
1790s	William Smith 1769–1839. Biostratigraphy: Smith establishes the principle of faunal succession while working as a surveyor/engineer for canal and mining projects.
1802	Reverend William Paley's *Natural Theology* defends the argument for design.
1815	William Smith's geological map of England is finally published.
1820	Jean-Baptiste Lamarck debates evolution and extinction with Cuvier.
1823	William Buckland 1784–1856 writes *Reliquiae Diluvianae* (relics of the flood).
1824	Megalasaurus is the first dinosaur fossil, identified and described by William Buckland.
1830	Etienne Geoffroy St. Hilaire defends a unity of plan connecting all animals in a debate with Cuvier.
1830–1833	Charles Lyell uses Etna as an example of the immense age of the earth and the value of uniformitarian methods in his *Principles of Geology*.
1831	Adam Sedgwick recants diluvialism, ending serious scientific attempts to link geology with biblical interpretation.
1831–1835	Charles Darwin travels on the *Beagle*.
1834	Jean de Charpentier argues carefully for the view that glaciers were once far more extensive than they are now.
1837	Darwin accepts evolution.
1837	Louis Agassiz accepts de Charpentier's glacial hypothesis and extends it to the ice age (*die Eiszeit*), becoming its greatest promoter.
1838	Darwin, inspired by Malthus, adopts natural selection as the mechanism for evolution.
1839	Roderick Murchison proposes the early Paleozoic Silurian system based on formations in Wales.
1842	Darwin writes a first sketch of his theory of evolution by natural selection.
1844	Robert Chambers publishes *Vestiges of the Natural History of Creation*, a popular but controversial book defending evolution.
1844	Darwin rewrites and extends his sketch of 1842, showing it to Emma for the first time.
1843	Richard Owen defines the word *homology* as "the same organ in different animals under every variety of form and function."
1852	Henry Walter Bates proposes Batesian mimicry.

1855	Alfred Russell Wallace proposes Wallace's law: Species appear near in both time and space to similar species.
1856	First Neanderthal fossils discovered.
1858	Wallace independently proposes natural selection, sending a letter on his ideas to Darwin.
1858	Darwin and Wallace's joint paper read to the Linnean Society that summer.
1859	*The Origin of Species* is published.
1859	Pasteur's jar experiments undermine spontaneous generation.
1860	Kelvin calculates the age of the earth at 40 to 400 million years, based on observations of steady cooling with no source of replacement heat, and obtains similar results for the sun.
1863	Gregor Mendel conducts breeding experiments with pea plants 1856 to 1863, publishes his particulate model of inheritance in 1863.
1870	Wide acceptance of evolution; natural selection remains controversial.
1870	John Tyndall publishes his work with the Tyndall box, leaving spontaneous generation untenable.
1871	Darwin suggests his so-called warm pond account of abiogenesis in a letter to Hooker.
1891–1894	Eugène Dubois discovers *Pithecanthropus erectus* (now known as *Homo erectus*) on the Island of Java.
1900	Hugo De Vries publishes his rediscovery of particulate inheritance; Mendel's earlier work is acknowledged.
1903	Pierre Currie measures the heat produced by radium. George Darwin and Joly point out the implications for the age of the earth.
1908	T. H. Morgan begins work on the genetics of fruit flies. Mutants are identified, in increasing numbers, from 1910 on, and their patterns of inheritance are investigated.
1912	Parts of a human-looking skull together with an ape-like jawbone are discovered in a gravel pit in Piltdown, England; the find is named *Piltdown man.*
1913	Different isotopes of the same element first distinguished, illuminating the process of radioactive decay.
1919	William E. Castle and Sewall Wright demonstrate that some genes influence the expression of other genes.
1925	Robert Dart discovers the Taung skull, naming it *Australopithecus africanus.*
1926	Arthur Holmes and others produce a report on radiological dating and the age of the earth. Reliability of these methods is now widely accepted.

1929–1937	Fossils of *Homo erectus* (the so-called Peking man) found near Beijing, China.
1930	R. A. Fisher publishes *The Genetical Theory of Natural Selection*.
1932	J.B.S. Haldane publishes *The Causes of Evolution;* also writes a series of papers on the mathematics of evolution published between 1924 and 1934.
1931	Sewall Wright presents his approach to unifying genetics and natural selection, including genetic drift as a distinct mechanism for evolution.
1936	Aleksandr Oparin publishes *The Origin of Life on Earth,* proposing that the building blocks of life were produced in an early, non-reducing atmosphere. J. B. S. Haldane expresses similar ideas.
1937–1944	Theodozius Dobzhansky, Ernst Mayr and G. G. Simpson apply the new synthesis of genetics and natural selection to field biology, taxonomy, and paleontology.
1953	Francis Crick and James Watson (drawing on Rosalind Franklin's X-ray crystallography work) establish the double-helix structure of DNA.
1953	Miller-Urey experiment. Amino acids and other organic molecules produced in a reducing atmosphere.
1953	The so-called Piltdown man shown to be a fraud produced from a modern human skull and the jaw of an orangutan.
1955	DNA polymerase discovered.
1960	First fossil of *Homo habilis* discovered in Olduvai Gorge, Tanzania.
1961–1968	The DNA code is cracked by Marshall Nirenberg, Har Gorbind Khorana, and Robert W. Holley.
1961	Jacques Monod and Francois Jacob publish 'The Regulation of Gene Expression' showing how lactase production is triggered.
1967	The RNA world is proposed by Carl Woese and named by Gilbert in 1968; some similar ideas go back even further.
1968	Motoo Kimura argues most genetic change at the molecular level is due to neutral mutations that spread (or disappear) by chance.
1975	Fred Sanger develops chain-terminator sequencing of DNA.
1977	Black smokers first discovered on the ocean floor, near the Galapagos Islands.
1982	First production of a human protein (insulin) using genetically modified bacteria. Later, HGH, GCSF, and others.
1983	PCR amplification of sample DNA invented by Kary Mullis.

1983   The homeobox (a highly conserved DNA sequence shared by DNA-regulating genes) is discovered.

1986   First genetically engineered vaccine, for hepatitis B.

1988   Günter Wächtershäuser proposes a metabolism-first model of abiogenesis, near black smokers.

1994   Flavr Savr Tomato is the first genetically engineered food on the market.

1994   *Ardipithecus ramidus* named; fossils of this species are later found to date back as far as 5.8 million years ago.

2000   The human genome project completes the first full sequence; others confirm relationships between of closely related species, including humans and chimpanzees.

2001   *Orrorin tugenensis* named based on fossils found in Kenya, dating to about 6 million years.

2002   *Sahelanthropus tchadensis* named based on fossils found in central Chad, dating to between 6 and 7 million years ago.

2003   *Homo floresiensis* named, based on fossils from the Indonesian island of Flores.

GLOSSARY

adaptation: A process in which a population comes to improve, on average, in how well it performs some function, or copes with some aspect of its environment. Also, a feature or trait that causes the improvement and is produced by the process.

allele: One form of a particular gene, for example, the blue allele for eye color.

altruism: Behavior on the part of one organism that helps another at a cost to the first.

animal: An organism belonging to the animal kingdom. Multicellular heterotrophs that (at least at some point in their life cycle) can move freely in their environment.

antennapedia: A mutation first noted in fruit flies, in which the fly's antennae are replaced by legs.

artificial selection: Any process in which plants or animals are deliberately bred to have some trait by keeping individuals that lack the trait from breeding.

autotroph: An organism that gets its nourishment (in particular, carbon) from inorganic sources. Includes most plants.

Batesian mimicry: A form of mimicry in which a species or variety is protected from predators because it resembles an unpalatable species.

biology: The study of living things.

biostratigraphy: The use of fossils to correlate layers of rock from different regions.

bithorax: A mutation in fruit flies which turns their normally tiny hind wings (which serve as balance organs) into full-sized wings.

botany: The study of plants.

bottleneck: A drastic temporary reduction in the size of a population.

carbohydrates: A class of chemicals based on carbon, hydrogen, and oxygen, including sugars, starches, and cellulose.

carnivore: An animal whose diet consists mainly of meat.

catastrophism: An approach to geology advocated by Georges Cuvier and widely accepted in the early nineteenth century. Catastrophists held that mountain ranges, valleys, and other large-scale geological features were produced by extremely violent events, unlike anything occurring in human history. Major floods were one important class of these; Cuvier held that such floods had wiped out all living things in large regions.

chain-terminator: A nucleobase altered so that, when it is added to a DNA molecule that is being built by polymerase, the chain cannot be extended any further.

clade: A group of organisms descended from a common ancestor.

class: A taxonomic rank, grouping orders or sub-classes, and grouped into a phylum.

competition: Any interaction between organisms in which a resource used by one is unavailable to another.

correlation: An association between two traits or properties. Two traits are positively correlated when an organism that has one trait is also more likely to have the other; they are negatively correlated when the opposite is true. Also (in stratigraphy), determining the time-relations between layers of rock (especially in different regions).

creationism: The view that existing species (or perhaps some higher taxon, such as genera) were created in their present form. Proponents of creationism today usually hold that the process of creation was supernatural.

diluvialism: Early- to mid-nineteenth century British form of catastrophism; these geologists held that so-called superficial gravels and related features were produced by the biblical (Noachian) flood.

diploblast: Any animal that forms two germ layers at gastrulation. Includes jellyfish and corals (*Cnidaria*), along with the comb jellies (*Ctenophores*).

domain (superkingdom): The highest level taxon. As proposed by Carl Woese, Domains divide all life into three basic groups: Bacteria, eukaryotes (organisms whose cells have nuclei), and archaea (a biochemically distinct branch of organisms superficially resembling bacteria)

DNA (deoxyribonucleic acid): The molecule that encodes our inherited traits; DNA is normally structured as a double helix; each side of the helix contains the information required to build a copy of the entire molecule.

drift (genetic drift): Changes in the gene pool of a population that are due to chance variations in which genes are passed from one generation to another, rather than selection for any advantage that some gene has over another.

Drosophila melanogaster: An important laboratory animal for genetic studies, famously used by T. H. Morgan in many experiments, and still widely used.

epigenesis: The process by which the context of cellular apparatus and the environment affects how genetic information is expressed during development. Also (older) a view of development holding that organisms grow from unformed matter, acquiring their form gradually in response to the surrounding environment. See *preformationism*.

enzyme: A protein that serves as a *catalyst*, speeding up some chemical reactions.

family: A taxonomic rank grouping tribes or genera together. Families in turn are grouped into orders.

fossil: Any trace of ancient life, including actual remains (sometimes modified by the replacement of the original materials with various minerals), tracks and even fossilized excrement (coprolites).

fungus (pl. fungi): An organism belonging to the kingdom fungi. A group of multicellular heterotrophic organisms that digest their food externally, absorbing the nutrients produced.

gastrulation: The point in animal development at which the embryo's cells migrate to form either germ layers: One in sponges, two in diploblasts, or three in triploblasts. All the animals organs develop from these layers.

geology: The study of the earth and its history.

gene: Originally an inheritable particle or element associated with some trait of the organisms that carry it. Now sometimes refers to a region of DNA that is translated to form a protein, or more broadly any region of DNA that influences when and where protein-producing genes are expressed.

gene pool: The range of alleles (variant forms of genes) that appear in a species or population. May include the distribution (i.e., frequencies) of these alleles.

genotype: The genes possessed by an organism (see *phenotype*); sometimes organisms with different genotypes may have the same phenotype—the difference in genotype can be revealed in offspring.

genus (pl. genera): A taxonomic rank grouping species together, genera are grouped in turn into tribes or families.

heterotroph: An organism that needs organic material for nourishment (specifically as a source of carbon). Includes both animals and fungi.

heterozygote: An organism having two different alleles of a particular gene of interest.

homeobox: A highly-conserved sequence of DNA appearing in Hox genes. The corresponding sequence of amino acids binds to DNA; genes containing homeoboxes act as switches regulating the expression of other genes.

hominin: Any organism on the human side of the divide separating the lineages of *Homo sapiens* and chimpanzees (the genus Pan). A member of the subtribe Hominina.

homology: A rich, multi-level similarity between organisms, inherited from a common ancestor. (See also *serial homology*.)

homozygote: An organism having identical alleles of a particular gene of interest.

hormone: A molecule that serves as a chemical messenger from one cell or group of cells to another. Hormones regulate and coordinate many activities in multi-cellular organisms.

Hox genes: A group of genes that serve as high-level switches controlling the development of different segments of an animal's body. Almost always found in the same head-to-tail order in the genome.

ice age: A period of time when the extent of continental glaciation is much wider than today's; more narrowly, the most recent period of widespread glaciation.

igneous rock: Any rock formed when magma (molten rock) solidifies.

intelligent design: A recent, minimalist form of creationism. Proponents of intelligent design hold that some sort of intelligent intervention is required for life to originate and/or evolve. However, unlike young earth creationists they do not all reject standard figures for the age of the earth, and unlike traditional creationists they do not always claim that different species or higher taxa were separately created.

kingdom: A taxonomic rank, grouping phyla together and in turn grouped into a superkingdom.

lipids: A class of organic molecules including fats (triglycerides), used in cell membranes and other places, largely made of carbon and hydrogen.

mammal: Any animal belonging to the mammalia class. Mammals have a single lower jaw bone (reptiles have three); they also have hair and suckle their young.

metamorphic rock: A rock which at some time since its formation was heated enough to cause its minerals to re-crystallize, but not enough to cause it to melt.

mycology: The study of fungi.

natural selection: The process proposed by Charles Darwin to explain how organisms come to be and stay adapted to their conditions of life as variations that make survival and reproduction more likely tend to spread through a population over time.

natural theology: The use of evidence about the natural world to support religious conclusions.

nucleic acids: RNA and DNA.

nucleotides: The units whose order, in a molecule of DNA or RNA, encodes genetic information. Nucleotides also have other roles in biochemistry—for example, adenosine triphosphate is the main so-called energy currency of cells.

order: A taxonomic rank grouping families together. Orders in turn are grouped into a sub-class or class.

peptide: A short sequence of amino acids linked together in a chain.

phenotype: The observable physical traits of an organism.

phylum: A taxonomic rank, grouping classes together and grouped into a kingdom.

polymerase: An enzyme that catalyzes the re-building of a double-stranded DNA from a single-strand.

polymerase chain reaction (PCR): One of the key tools of molecular genetics, this reaction is used to amplify a sample of DNA by producing many copies of a selected sequence in the sample.

polypeptide: See *protein*.

population: A group of organisms of the same species living in a particular region.

precipitation: A chemical process in which dissolved material comes out of solution, forming a solid.

preformationism: The doctrine that organisms develop from matter that already has form. In its most extreme version, preformationism held that embryos are complete organisms in miniature and their development requires only growth. See *epigenesis*.

protein: A sequence of amino acids longer than a peptide. One of the fundamental families of organic molecules.

radiological dating: Any technique for dating materials that relies on the rate of decay of some radioactive isotope.

reducing: A chemical environment lacking (or low in) free oxygen.

RNA (ribonucleic acid): A nucleic acid used by all life in many processes, including the building of proteins and controlling the expression of various genes. Also used to store genetic information in some viruses.

sedimentary rock: A rock formed from clasts (small particles of pre-existing rock or organic material such as shells) cemented together, or by precipitation from a solution.

sequencing: A chemical procedure for determining the order of nucleobases in some nucleic acid or of amino acids in some peptide or protein.

serial homology: A rich, multi-level similarity between different parts of the same organism, produced by the same genes (or genes modified from a common ancestral gene).

sexual selection: An aspect of natural selection first proposed by Charles Darwin, in which traits that increase an organism's chances successfully obtaining a mate are selected for.

species: The lowest taxonomic rank. Groups individual organisms (also sometimes varieties or sub-species) and grouped in turn into a genus.

stratigraphy: The study of the layered structure of rocks forming the earth's crust.

taxonomy: A system of kinds into which we group some collection of things; in biology, the system of kinds we apply to group organisms.

taxon: A group of organisms gathered together at some taxonomic rank.

taxonomic rank: A level in a hierarchical system of taxonomy.

tribe: A taxonomic rank sometimes inserted between genera and a family, gathering the genera together. Tribes in turn are grouped into families.

triploblast: Any animal that goes through a developmental phase (called gastrulation), in which three layers of tissue (ectoderm, endoderm, and mesoderm) are formed. Includes all the intermediate and higher animals, from flatworms to vertebrates.

uniformitarianism: In geology, the view that present processes, acting over time, are sufficient to explain the history of the earth.

varve: An annual layer of sediment or sedimentary rock. These layers can be distinguished because of seasonal variation in the size and rate at which particles are laid down.

young earth creationism: A view of geology and biology based on a literal reading of Genesis, including separate creation of different species (or some slightly higher taxons) and a roughly 10 thousand year history for the earth.

zoology: The study of animals.

BIBLIOGRAPHY

Albritton, Claude. 1980. *The Abyss of Time.* San Francisco: Freeman, Cooper. A reflective history of geology from Steno to the twentieth century.

Barrett, P. H. 1974. Early writings of Charles Darwin. In Gruber, H. E., *Darwin on man. A psychological study of scientific creativity; together with Darwin's early and unpublished notebooks. Transcribed and annotated by Paul H. Barrett, commentary by Howard E. Gruber. Foreword by Jean Piaget.* London: Wildwood House.

Buffon, G.L. 1749–1804. *Histoire Naturelle, Générale et Particulière.* Paris: Imprimerie Royale, Puis Plassan.

Burchfield, Joe D. 1975. *Lord Kelvin and the Age of the Earth.* London: MacMillan. A detailed account of the debate over the age of the earth in the late nineteenth century and how it was finally resolved.

Carroll, Sean B. 2005. *Endless Forms Most Beautiful: The New Science of Evo-Devo.* New York: W.W. Norton. An insider's guide to animal development, a field that's growing by leaps and bounds as the tools of modern biochemistry illuminate (sometimes literally) the signals that control development, and the implications of these new discoveries for evolution.

Cutler, Alan. 2003. *The Seashell on the Mountaintop.* New York: Dutton. A history of early geological thinking, centered on the story of Nicolaus Steno.

Darwin, Charles. 1839. *The Voyage of the Beagle.* London: Henry Colburn. A wonderful Victorian traveler's story. Like all of Darwin's works, this book has been re-printed more than once, and is available on the web at Darwin-online.org.uk.

———. 1859. *On the Origin of Species by Means of Natural Selection, or the Preservation of Favoured Races in the Struggle for Existence.* London: John Murray. The original case for evolution by natural selection, a scientific classic.

————. 1871. *The Descent of Man and Selection in Relation to Sex.* London: John Murray. In this book Darwin finally addresses the issues about human ancestry raised by his theory of evolution.

————. 1872. *The Expression of the Emotions in Man and Animal.* London: John Murray. A study of the widely shared expressions and patterns of behavior that express emotions.

Darwin, Erasmus. 1796. *Zoonomia; or, The Laws of Organic Life.* Parts I-III. London: J. Johnson

Darwin, Francis. 1902. *The Life of Charles Darwin.* London: John Murray (Republished in 1995. London: Studio Editions Ltd.). A classic biography, written by Darwin's son. Includes extensive material from his correspondence.

Darwin, Francis (ed.). 1887. *The life and letters of Charles Darwin, including an autobiographical chapter.* London: John Murray. Volume 1.

Dawkins, Richard. 1976. *The Selfish Gene.* Oxford and New York: Oxford University Press. A defense of Dawkins's gene-oriented approach to understanding evolution.

————. 1996. *Climbing Mount Improbable.* London: W.W. Norton & Company. A clear and convincing explanation of how natural selection can produce complex and sophisticated adaptations.

Dennett, Daniel. 1995. *Darwin's Dangerous Idea: Evolution and the Meanings of Life.* New York: Simon and Schuster. Dennett argues that only natural selection can provide a real explanation of adaptation, criticizing both biologists who have questioned the power of natural selection to produce optimal adaptations and scholars outside biology who have not recognized the importance of natural selection for their own specialties.

de Waal, Franz. 1996. *Good Natured: The Origins of Right and Wrong in Humans and Other Animals.* Cambridge, MA: Harvard University Press. De Waal examines altruism, empathy, and cooperative behavior in some of our nearest relatives.

Eve A.S. 1939. *Rutherford.* Being the Life and Letters of the Rt. Hon. Lord Rutherford, O. M.: New York: Macmillan.

Fisher, R. A.,1930. *The Genetical Theory of Natural Selection.* Oxford: Clarendon Press. (Republished in 1999. New York: Oxford University Press.) The first book-length presentation of the modern synthesis of Mendelian genetics with natural selection, Fisher applies his expertise in statistical analysis to population genetics.

Forest, Barbara. 2004. *Creationism's Trojan Horse: The Wedge of Intelligent Design.* Oxford: Oxford University Press. A critical examination of the so-called intelligent design movement.

Fortey, Richard. 1998. *Life: A Natural History of the First Four Billion Years of Life on Earth.* New York: Knopf. The history of life, ranging from evidence of the very earliest living things to the origins of *Homo sapiens.*

————. 2000. *Trilobite! Eyewitness to Evolution.* London: Harper-Collins. Everything you ever wanted to know about trilobites: their history, their

different forms, their spectacular eyes, from a paleontologist who special-
izes in these strange Paleozoic arthropods.

————. 2004. *The Earth, An Intimate History*. London: Harper. An introduction to geology today, structured around personal travels in places where
our understanding of the earth makes contact with the everyday world
around us.

Futuyma, Douglas. 1995. *Science on Trial: The Case for Evolution*. 2nd ed.
Sunderland, MA: Sinauer Press. A vigorous defense of evolution, presenting evidence from taxonomy, paleontology, biochemistry, and other sources
along with reflections on the philosophy of science along with a strong critique of so-called scientific creationism.

————. 1998. *Evolutionary Biology*. 3rd ed. Sunderland, MA: Sinauer Press.
A standard evolutionary biology text book, and an excellent place to begin
a serious study of evolution.

Gohau, Gabriel. 1990. *A History of Geology*. Rutgers: Rutgers University Press.
A fascinating and accessible account of the history of geology, by an expert
in the field. (Translated from the French by A. V. Carrozi and M. Carrozi.)

Gosse, Edmund. 1907. *Father and Son*. London: William Hyndeman Ltd. (Re-
printed in 1989. New York: Penguin Books.) An autobiography focusing
on Edmund Gosse's relationship with his father, Philip Gosse, who was a
zoologist, fundamentalist Christian, and the author of *Omphalos*, in which
the elder Gosse argued that the apparent geological age of the earth was
an illusion built into the planet at its creation.

Gosse, Philip. 1857. *Omphalos: An Attempt to Untie the Geological Knot*.
London: John Van Voorst. (Reprinted in 1998. Woodbridge, CT: Ox Bow
Press.) Gosse proposes to reconcile a literal reading of Genesis with the
sciences of geology and paleontology by invoking the hypothesis of a so-
called false past, built into the world at its creation.

Gould, Stephen J. 1977. *Ontogeny and Phylogeny*. Cambridge, MA: Harvard
University Press. A detailed historical study focusing on nineteenth century debates over development, with remarks on contemporary issues
concerning evolution, and accelerated and slowed development.

————. 1987. *Time's Arrow/Time's Cycle: Myth and Metaphor in the Discovery
of Geological Time*. Cambridge, MA: Harvard University Press. An exploration of large-scale ideas about geological time.

————. 2002. *The Structure of Evolutionary Theory*. Cambridge, MA: Harvard
University Press. A massive work presenting Gould's views on evolution,
emphasizing punctuated equilibrium, the balance between natural selection and non-adaptive causes and constraints in evolution, and the importance of catastrophic and unpredictable events like the impact that ended
the Cretaceous.

Haldane, J. B. S. "A Mathematical Theory of Natural and Artificial Selection."
A series of papers published between 1924 and 1934, in the *Proceedings*
and the *Transactions of the Cambridge Philosophical Society* and in *Genet-*

ics. Haldane presents the results of his work combining statistics, Mendelian genetics, and selection.

————. 1932. *The Causes of Evolution*. London: Longman's, Green and Co. (Republished in 1990. Princeton: The Princeton Science Library, Princeton University Press.) One of the founders of the modern synthesis presents his account of evolution.

Hallam, Anthony. 1989. *Great Geological Controversies* (2nd ed.) Oxford: Oxford University Press. A history of geology centered on a series of controversies: Neptunism vs. Hutton; catastrophism vs. actualism vs. uniformitarianism; the Ice Age; the age of the earth; continental drift.

Hazen, Robert M. 2005. *Gen-e-sis, The Scientific Quest for Life's Origin*. Washington: Joseph Henry Press. An up-to-date, non-technical introduction to research on abiogenesis today, told by an active researcher.

Hutton, James. 1788. *Theory of the Earth, in Transactions of the Royal Society of Edinburgh*, vol. I, Part II, pp. 209–304.

Imbrie, John and Palmer Imbrie, Katherine. 1979. *Ice Ages: Solving the Mystery*. Cambridge, MA: Harvard University Press. A historical examination of ice ages and how we came to understand their causes.

Keynes, Randall. 2001. *Annie's Box: Charles Darwin, his Daughter and Human Evolution*. London: Harper Collins. A personal history, by a great-grandson of Charles Darwin, focusing on Darwin's relationship with his daughter Annie, her death, and Darwin's feelings about evolution, God, and humanity's place in the natural world.

Kitcher, Philip. 1982. *Abusing Science: The Case Against Creationism*. Cambridge, MA: MIT Press. A philosopher of science dissects the so-called scientific creationism movement.

Kimura, Motoo. 1983. *The Neutral Theory of Molecular Evolution*. Cambridge: Cambridge University Press. Presents Kimura's influential (1968) argument for the importance of unselected, chance shifts in the frequency of alleles in molecular evolution.

Lane, N. Gary. 1992. *Life of the Past* (3rd ed.). New York: Macmillan. An overview of the history of life by a professional paleontologist.

Leibniz, G.W. 1690-91(?). *Protogaea*. A manuscript first published in a 1749 collection of Leibniz' works. *Svmmi polyhistoris, Godefridi Gvilielmi Leibnitii Protogaea, sive de Prima Facie Tellvris et Antiqvissimae Historiae Vestigiis in ipsis Natvrae Monvmentis Dissertatio ex Schedis Manvscriptis Viri Illvstris in Lvcem Edita Christiano Lvdovico Scheido*, Goettingae: Svmptibvs Ioh. Gvil Schmidii, Bibliopolae Vniversit., 1749. xxviii+86pp. Available in English at http://www.leibniz-translations.com/pdf/protogaea.pdf .

Lyell, Charles. 1830–33. *Principles of Geology*. London: John Murray. A geological classic; Lyell puts his case for uniformitarianism. An important influence for Darwin.

Majerus, M. E. N. 1998. *Melanism: Evolution in Action*. Oxford: Oxford University Press. A detailed examination of H. B. D. Kettlewell's experiments and subsequent work on industrial melanism. Majerus rejects recent creationist claims that Kettlewell's experiments were faked or misleading.

Mayr, Ernst. 1942. *Systematics and the Origin of Species from the Viewpoint of a Zoologist*. New York. Columbia University Press. (Reprinted in 1982 with an introduction by Niles Eldridge.) A classic of evolutionary biology, presenting Mayr's fieldwork and his biological species concept. Mayr continued to write widely on biology and the history and philosophy of biology until his death in 2005, at the age of 100.

———. 1982. *The growth of biological thought*. Harvard University Press, Cambridge, MA.

McPhee, John. 1998. *Annals of the Former World*. New York. Farrar, Strauss, and Giroux. Gathers together McPhee's wonderful books on geology, based on travels with geologists across the United States.

Ospovat, Dov, 1981. *The Development of Darwin's Theory: Natural History, Natural Theology and Natural Selection 1838–1859*. Cambridge and New York: Cambridge University Press. A detailed history of Darwin's thinking on natural theology, variation, and adaptation.

Owen, Richard. 1843. *Lectures on the Comparative Anatomy and Physiology of the Invertebrate Animals*. London : Longman, Brown, Green, and Longmans.

Paley, William. 1802. *Natural Theology: Evidence of the Existence and Attributes of the Deity, Collected from the Appearance of Nature*. The standard version of the argument for design, including the famous watch analogy; an important influence on Darwin.

Pennock, Robert T. 1999. *Tower of Babel: The Evidence Against the New Creationism*. Cambridge, MA: MIT Press. A detailed critique of intelligent design creationism.

——— (ed.). 2001. *Intelligent Design Creationism and Its Critics*. Cambridge, MA, London: MIT Press. A collection of papers from critics and defenders of intelligent design creationism.

Pope, Alexander. 1734. *An Essay on Man, Being the First Book of Ethic Epistles. To Henry St. John L. Bolingbroke*. [Epistles I-IV] London: Printed by John Wright for Lawton Gilliver.

Quammen, David. 1996. *The Song of the Dodo: Island Biogeography in an Age of Extinctions*. New York: Scribner. Life depends on other living things, and is often threatened by other living things. To be isolated is to be both vulnerable and protected. Historically, life on islands has suffered more and faster extinctions than life on the mainland with the arrival of human beings (and their companion forms of life). Today's increasing rate of extinctions worldwide is partly due to the division of wild habitat into small, isolated areas.

Rabinow, Paul. 1996. *Making PCR, A Story of Biotechnology*. Chicago: University of Chicago Press. A detailed history of the companies and personalities involved in the development of PCR (the polymerase chain reaction).

Raby, Peter. 1996. *Bright Paradise, Victorian Scientific Travellers*, London: Random House. A history of the nineteenth century scientific travelers, including von Humbolt, Darwin, Wallace, and Bates, who did so much to expand our knowledge of life around the world.

Rudwick, Martin J. S., *The Meaning of Fossils: Episodes in the History of Paleontology*. 2nd ed., New York: Science History Publications. A rich account of the history of ideas about fossils since the sixteenth century.

Ruse, Michael. 1981. *Darwinism Defended: A guide to the evolution controversies*. Reading, MA and London: Addison-Wesley. An energetic account of Darwin's theory, its history, and the creationist controversy in the United States. Also examines sociobiology and relations between evolution, natural selection, and ethics.

———. 2003. *Darwin and Design: Does Evolution Have a Purpose?* Cambridge, MA: Harvard University Press. An exploration of the place of teleology in evolution.

———. 2005. *The Evolution-Creation Struggle*. Cambridge, MA: Harvard University Press. An exploration of the religious impulses and ideological tensions that have made evolution so controversial in the public sphere (especially in the United States) even as it triumphed as a scientific theory.

Russell, Michael. 2006. "First Life," *American Scientist* 94(1) 32. Presents Russell's work with Martin and others on hot springs, metabolic pathways, and the origins of life.

Simpson, George Gaylord. 1944. *Tempo and Mode in Evolution*. New York: Columbia University Press. Simpson extends the modern synthesis, applying it to paleontology.

———. 1951. *Horses*. New York: Oxford University Press. In this book Simpson details the evolutionary history of horses, showing that it was not a simple march from *Eocene hyracotherium* to *Equus* (the modern horses).

Sober, Eliot. 1984. *The Nature of Selection*. Cambridge, MA: Bradford Books. Sober is a philosopher specializing in the philosophy of biology. His book is a comprehensive study of the conceptual structure of evolutionary theory.

Sober, Eliot and Wilson, David. 1998. *Unto Others*. Cambridge, MA: Harvard University Press. An engaging philosophical exploration of the evolution of altruism.

Southwood, Richard. 2003. *The Story of Life*. Oxford: Oxford University Press. A history of life on earth from the beginning to the present, with discussion of the origins of life, mass extinctions, and human evolution.

Tattersall, Ian. 1998. *Becoming Human, Evolution and Human Understanding*. New York: Harcourt Brace and Company. Reflections on the human fossil and archeological record.

Wallace, Alfred Russel. 1855. On the law which has regulated the Introduction of New Species. Annals and Magazine of Natural History, including Zoology, Botany, and Geology 16: (September): 184-196.

Weiner, Jonathan. 1995. *The Beak of the Finch: A Story of Evolution in Our Time.* New York: Vintage Books. The story of some recent research on evolutionary changes in the beaks of Galapagos finches, changes that can be observed today as short-term shifts in climate alter food supplies.

Wilberforce, Samuel. 1874. *Essays Contributed to the Quarterly Review.* London: J. Murray.

Wills, Christopher and Bada, Jeffrey. 2000. *The Spark of Life.* Cambridge, MA: Perseus. An account of research on the origins of life.

Wood, Bernard. 2005. *Human Evolution: A Very Short Introduction.* Oxford: Oxford University Press. A very readable and up-to-date handbook on the evolutionary history of our species.

Wright, Sewall. 1931. "Evolution in Mendelian Populations," *Genetics* 16: 97–159.

———. 1932 "The roles of mutation, inbreeding, crossbreeding and selection in evolution," *Proc. 6th Int. Cong. Genet.* 1: 356–366.

Young, David. 1992. *The Discovery of Evolution,* Cambridge: Natural History Museum Publications and Cambridge University Press. A beautifully illustrated history of evolution, from taxonomic developments in the middle of the seventeenth century up to the modern synthesis.

INDEX

ABOUT THE AUTHOR

Bryson Brown is professor of philosophy at the University of Lethbridge. Born in Niagara Falls, Ontario, he received his PhD in philosophy from the University of Pittsburgh in 1985. Though his research focuses chiefly on logic and philosophy of science, he also teaches a course covering the history of geology and biology, which he has always loved.